Heating Your Home with Wood

by Neil Soderstrom

Drawings by
Vantage Art, Inc.

POPULAR SCIENCE
HARPER & ROW
New York, Evanston, San Francisco, London

To
my mom and dad
and
my wife Hannelore —
and
to the memory of
Carlos Rodriguez

Contents

Preface

"I DON'T BELIEVE IT!" exclaimed our oil man over the phone. At my request, he was examining our record of fuel consumption over the several years we'd owned the house. He verified that, on average, our annual oil needs had been only one-fourth that of the previous owners. Yet our most recent winters had been the coldest in many years. Of course, there is a reason we did better than the previous owners. They relied solely on an oil burner. We added a wood stove and a little insulation.

It is late May as I write this. We live in New York State and keep the main-floor thermostat set at 65°F. Yet our oil burner hasn't fired since February. Our little 115-pound Scandinavian stove can handle the entire heating load for our 1500-square-foot house, except on the coldest winter days. Even then, the oil burner fires briefly only a few times a day. Oh, we'd have used far less oil over the years if we hadn't run short of wood at times. Then too, we were often forced to burn wood that hadn't seasoned dry enough to yield nearly its full heating potential.

Wood heat has done more than cut our oil costs. It's made us feel more self-sufficient. It's made the home cozier than oil pumps and squirrel-cage blowers ever could. Along the way we've learned from the writings of veteran wood burners. We've exchanged notes with wood-burning neighbors and far-flung correspondents. We've also tested our own notions and some inventions with good results.

It is my hope that this book will help you approach wood heating well informed. Many people simply plunge into wood heating with only the knowledge they've gained over campfires. Of course, there's much to learn there. But wood heat as a modern energy alternative is now attracting scientists and engineers who can explain why some old wood-heating methods were good ones after all and why others weren't. We're learning more about the importance of house design, stove design, and the burning qualities of the various woods. We're learning how to prevent house fires, how to avoid insect infestations, and how to harvest wood without ruining the landscape or maiming ourselves in the process. More on those aspects later.

Many people contributed to this book in noteworthy ways. My thanks go first to the publishers: the people at the Book Division of Times Mirror Magazines, Inc. To John Sill, who more than once has offered me grand opportunities and then stuck by with encouragement and contributions. To Adele Bowers, supporter of energy-conservation programs and this book project in particular. To Henry Gross, who supervised the book's editing and art development, while considerately dealing necessary light touches and full nelsons. To Jeff Fitschen and Pat Blair, colleagues who helped with art and editorial matters while mending loose ends elsewhere without a whimper. To Moya Sullivan, faithful supplier of wood-heat lore on subjects as diverse as power splitters and carpenter ants. To Larry de Quintal, who started me off some years ago with a good stove. To Ted Johnson, owner of three wood stoves, who copyedited the manuscript. To Donna Baier, who proofed the galleys. To Andy Steigmeier, who designed the pages. And to Phil Kennedy, who provided the artwork.

Many nationally recognized authorities on wood heat lent support too. First among these was physics professor Jay Shelton, author of *The Woodburner's Encyclopedia*. I leaned heavily on his scholarly book for an understanding of burning principles and then benefited from his comments on my early chapters. Dr. Shelton's encyclopedia and other publications resulting from his ongoing research are available from him directly % Physics Department, Williams College, Williamstown, MA 01267.

Thanks also to Larry Gay, chemist, stove maker, and author of *The Complete Book of Heating with Wood*, a pioneer book in this field. Mr. Gay gave invaluable suggestions for my stoves chapter and voiced important cautions elsewhere.

Thanks to Blake Stretton, former gyppo logger and now a vice president of Stihl, Inc., who improved the chapters on chain saws and logging. To D. Douglas Dent, timber falling consultant, of Beaverton, Oregon, whose book *Professional Timber Falling* and loan of rare publications came in handy. To Dr. John I. Zerbe, energy research manager with the U.S. Forest Products Laboratory, for vital help with the chapters on wood fuel and seasoning. To Daryl de Angelis of Monsanto and Dennis Murphy of Vail, Colorado, for insights into the air-pollution factors of wood heating. To Dr. Stanley Barras and Dr. John Peacock, entomologists with the Department of Agriculture, for guidance on the hazards of insect-infested wood. To Robert Boardman of the National Audubon Society, for thoughts on the effects of wood cutting in an ecosystem. To Frank Smith, veteran wood burner and woodsman, who commented on the manuscript and allowed us to photograph his woodpiles. And to A. J. Hand, an extraordinary writer-photographer, whose *Home Energy How-to* offered countless models for my text and illustrations.

Scores of state and provincial foresters and university extension foresters provided wood-burning profiles on their own regions. For efforts beyond what was asked of them, special thanks to the following: Calvin Kerr, Alaska; Tim French, Arkansas; John Shelly, California; William Wilcox, Colorado; Vernon Burlison, Idaho; T. W. Curtin, Illinois; Paul Wray, Iowa; Leonard Gould, Kansas; Donald McFatter, Louisiana; George Bourassa and Timothy O'Keefe, Maine; Ted Cady, Massachusetts; Paul Slusher, Missouri; Philip Verrier, New Hampshire; Kurtis Atkinson, Oklahoma; E. P. Farrand, Pennsylvania; David Holt, Rhode Island; Max Young, Tennessee; Michael Grosjean, Utah; E. Bradford Walker, Vermont; Harry Thorne, Wisconsin; Bob Lewis, New Brunswick; W. S. Bailey, Saskatchewan.

On the home front, thanks for contributions go to neighbors Tom Grundvig, Ken Orser, Clyde Teetsel, Carl Bruder, Paul Fisher, Bill Barlow, Dave Pettit, and Donna Williams. For sharing their wood-burning savvy over the years, thanks to my dad and to my brother Rob. Finally, for the most enormous help of all, thanks to my wife Hannelore and son Nikolai.

NEIL SODERSTROM
Shenorock, N.Y.

1 | Is Wood Heat For You?

Since the oil embargo of 1973–74, North Americans by the millions have turned to wood heat as a matter of precaution, economics, and taste. Scientists had long been forecasting an energy crisis, but it took the oil embargo to awaken most of us. Almost overnight, wood heat ceased to seem quaint and backward. Suddenly, it was the rage—the fashionable precaution, the way to be independent of fossil fuels and electric power. And it could be used for cooking as well.

Country stores that customarily sold a few wood stoves each winter were soon peddling more than 100 per week. City dwellers often paid dearly for wood, carrying armloads past their doorman before pushing the elevator button for the umpteenth floor. Suburbanites found that wood could be as cheap as fossil fuels for equivalent heating values. In rural areas, delivered wood sometimes arrived at neighborly rates. And people who could cut their own wood pronounced it free bounty from nature—aside from time, labor, and costs for equipment. So it seemed economically appealing too.

How about taste? A wood-heated home is a special place. It can be astoundingly cozy, the stove emanating a character of warmth that no fossil-fueled furnace can equal. There's that little stove turning out 30,000 to 50,000 Btu per hour, enough heat for a three- to six-room home in a northern climate. As it whispers through the winter it becomes the center of family life—the teapot always ready for hot drinks or dish water, the wet clothes drying after a day in the snow, the bread baking on top, the dog, finally unable to lie so close, moving to a cool corner.

Indeed, wood heat can be beautiful. None miss it more than the families who exhaust their wood supply before the winter is over. Suddenly, there's a sense of loss—perhaps even desolation—that no one felt before the wood stove was installed. The loss is beyond that of simple heat. It is undefinable. But it is there, and it weighs heavily. Everyone vows to have enough wood put by for the next winter. This is a vow of the sincerest kind. Yet it is only a vow.

REAL HEAT AND REAL COSTS. A well-built, well-insulated smaller home can be heated in northern climates on about five cords of wood each year. Especially well-designed homes that exploit the sun can manage on a cord or two each year. Larger homes may need ten cords. Large, drafty homes employing fireplaces may need twenty cords.

A cord of wood is a neat stack measuring 4×4×8 feet. Normally, a single hardwood tree with 20-inch base diameter will yield a cord of wood. Yet it may take 200 trees with 3-inch base diameters to give you the same amount of wood. The densest woods give off the most heat, simply because they have the greatest mass available for combustion. Per cord, dense woods such as hickory and oak

1

can yield as much real heat to the home as 175 gallons of fuel oil or nearly 25,000 cubic feet of natural gas. Lightweight woods, weighing half as much, will yield only half as much heat.

Normally, wood suppliers will deliver mixed cords that contain some woods of heavy, middle, and light weight. So you should usually expect to receive a cord that may equal only 120 gallons of fuel oil or about 18,000 cubic feet of natural gas. (That's provided your supplier gives you a full cord. More on that in Chapter 12.)

Okay, let's say you've figured that costs per cord and costs for fossil fuel show that wood is more economical. If the wood is delivered at stove length and split small enough to fit easily through the stove door alongside another log or two, your only further outlay is the labor for storing the wood outdoors, lugging it indoors, and maintaining the stove, stovepipe, and chimney. If you must saw and split whole logs, you'll have to figure in costs for hand tools, at least, plus your labor. If you decide to do your own logging, you'll probably need a chain saw and either a trailer or a pickup truck. If you already own some wooded acreage, you'll be in business. If not, you'll have to locate some wood and then get permission to take it. Logging and tree work are covered in Chapter 17.

Good wood stoves cost from $200 up. If you can use an existing chimney flue, other chief installation costs will be for stovepipe and heat-protective shields for walls, floor, and furniture. But if you'll need a new chimney, costs may run anywhere from $200 to more than $1000. Wood-burning furnaces and furnaces that burn both wood and another fuel cost anywhere from $600 to $2500. As for fireplaces, you'll find that they don't receive a hearty endorsement in this book—not as a means of heating, anyway.

Poorly operated fireplaces lose more heat than they add. Well-designed and well-operated fireplaces can contribute some heat, but the heating value in relation to the wood consumed and the air pollution still make them environmentally unreasonable—though lovely—relics from an age that sported few heating alternatives, far cleaner air, far greater wood supplies, and far fewer fireplaces. (The U.S. Department of Commerce reported an increase of firplaces in new house construction from 38 percent in 1972 to 58 percent in 1976! Here, presumably, consumers have shown—or advertisers have instilled—an increased, but questionable, interest in fireplaces as alternate means of heating. For atmosphere on a winter's evening, they are dandy. For heat, they are not.)

It is absolutely vital that your wood-burning system be a safe one as well as an efficient producer of heat. Cutting corners on installation costs can increase fire hazards and thus make the whole wood-heat proposition a mistake. Here you should work closely with local building officials regardless of your neighbor's or your stove salesman's corner-cutting tips.

LABOR AND TOOLS. Wood burning is a labor-intensive means of heating. If you buy wood already cut and split, your supplier may have to charge you at least $9 for every $1 he pays for the wood on the stump, or else he may not realize the federal hourly minimum wage. Sometimes, though, you can get especially good prices from tree-work and landscaping contractors who get paid for removals and and then use your backyard for the first dump—minimizing handling and your costs.

If you cut your own wood, and especially if you have to haul it any distance, you may be mortified if you try to justify your hourly outlay. And when you begin amortizing costs for tools and machinery, while tacking on annual costs for fuel, maintenance, and repairs, you may reprimand the muses that beguiled you into the work.

It's best not to try to justify the work in terms of money. Consider the fresh air and exercise. Be pleased with your increasing knowledge of trees. Take heart in your improving skills with tools. Feel the satisfaction of dropping a tree in tight quarters, just as you'd planned it. Realize your role as part-time forester. Know the effects that your logging will have on the landscape.

ENVIRONMENTAL IMPACTS. Wood is a renewable resource. Many parts of North America have enough trees and land for sustained yields of wood for the timber industry, as well as for goodly amounts of home heating. The final chapter in this book describes the many ways you can obtain wood without ruining the landscape. Many people gather cords of fuel wood each year without ever felling a living tree.

Yet dead and dying trees play a vital role in nature too. Without them, there would be fewer creepers and crawlers in a food-chain link of interest to bears and woodpeckers. Too much gleaning—overbrowsing—for wood can eliminate vital links in nature's food chain and also rob the soil of nutrients that would result from decay. Still, there are ways of harvesting wood selectively, without disrupting food chains or significantly depleting the soils.

How about air pollution? All fuels pollute the air when they are burned. No emissions researcher would recommend that you use the top of your chimney flue as an inhalator. Yet wood, for the most part, doesn't add the same pollutants to the air that petroleum-based fuels do. The effects of wood burning on air quality are covered at the end of the next chapter.

Can wood burning end the world energy crisis? Certainly not. Much about wood heat is good, even wonderful, and it helps diminish the consumption of fossil fuels. Solar and other technologies may someday be able to meet everyone's home heating needs. Yet even solar collectors will require enormous quantities of metals of limited supply. Then there's the pollution resulting from the manufacture of solar-heating products.

Let's hope that the world can tool up fast enough to avert catastrophic energy shortages without generating catastrophic amounts of pollution in the process. Of course the world would be better off with a stabilized and, yes, decreasing population. But that's for other books and stoveside chats.

Meanwhile, let's look at wood heat and the ways of making it an environmentally *reasonable* means of home heating, as well as a perennial delight.

2 | What Happens When Wood Burns?

MANY PEOPLE ARE masters at starting and maintaining a wood fire, yet don't really know what fire is. Then too, it's possible to understand fire as a phenomenon and still lack the knack for coaxing a fire from kindling and then banking it properly for the night. Yet for masters as well as beginners, a knowledge of the "lab science" of fire makes wood burning a surer proposition—and all the more fascinating.

COMBUSTION BASICS. Fire results from chemical reactions. The chemical energy stored in the wood is converted to heat and light, as well as to infrared radiation. Under the right conditions, once this conversion process starts, it builds upon itself, in a complex chain of reactions that release lots of energy and leave only a little ash.

Wood, like gas and oil, is made essentially of compounds of carbon, oxygen, and hydrogen. When wood is burned completely—perfectly oxidized—under laboratory conditions, its by-products are simply water vapor and carbon dioxide, the same stuff we exhale. But in an ordinary stove or fireplace, wood burning is seldom complete. The by-products of these incomplete burns are assorted flue gases and liquids, as well as solid particles.

To have a fire, you need fuel, heat, and oxygen. That's it. But these three ingredients must interact within ideal ranges or else the fire may fail to light, or if it lights, it may only flutter weakly and then die. You can start a wood fire without using a flame, if you apply enough heat. You can also heat wood far beyond its normal ignition temperature and still have no fire if you deny the wood sufficient combustion air. (Incidentally, heating a material in the absence of air is called pyrolysis; commercial charcoal is produced this way.) You can even heat combustion air enough for fire making, but you won't have fire unless the wood meets certain requirements for burning.

HOW WOOD BURNS. In the strictest sense, wood itself doesn't burn. Its gases burn, and its charcoal burns. But the wood itself does not. Here's how it all works:

Wood is essentially made of cellulose fibers and a bonding compound called lignin, as well as lesser compounds and lots of water. Again, these basic components of wood are made of just carbon, hydrogen, and oxygen. When heat is applied, it makes the molecular structures in wood become increasingly active. As the wood's surface temperature approaches 212°F (100°C), the water in the wood begins to boil and evaporates as steam. As long as the water remains in the wood, its boiling and evaporation rob heat energy from the source and thereby tend to keep the wood cells from gaining more heat. But if the temperature of the heat source is high enough and persistent enough, it boils out more and more water from the wood's inner layers.

As the wood surface temperature rises beyond 212°F to about 540°F, other gases and liquids are produced. In this temperature range the major gases are carbon dioxide, carbon monoxide, and the acetic and formic acids that are abundant in creosote. As these new gases are forming, they consume heat from the outside heat source, generating no heat themselves. Given the right air mixtures, carbon monoxide and the acids could ignite at these temperatures, but they are still too heavily mixed with noncombustibles to ignite.

A new phase begins once the outside heat source has raised the wood's surface above 540°F and continues to heat it toward 900°F. Now the reactions begin to produce their own heat and to generate gases, including methane (the basis of natural gas) and methanol (wood alcohol) as well as more acids, water vapor, and carbon dioxide. Tar droplets emerge and are borne upward by the gases. As these gases emerge from the wood, they mix with air and ignite at temperatures of about 1100°F. Once ignited, these gases burn at about 2000°F.

Ever since reaching 540°F, the surface of the wood has been drying, charring, withering. Yet even at 900°F, the surface charcoal is not hot enough to ignite the gases still emanating from within. For ignition, those gases depend on 1100°F heat supplied from other burning wood.

So far, we've looked at the wood-combustion process as though it runs a straight line. It doesn't. Although one small area on one piece of wood tends to progress predictably, various portions of that same piece of wood will be in different stages of the combustion process. For example, even though the surface of a piece of wood may have reached 900°F, the core may still be quite cold and moist. The core warms only as the wood heats it from the outside, layer by layer, liberating gases and liquids that form new compounds all the while.

THE FLAMES OF A FIRE. In some respects, a piece of wood burns much the way a candle does, except that wood involves a far more complex process. Heat from the candle flame melts the paraffin near the base of the wick. The melted paraffin travels up the wick. Melted paraffin on the wick and at its base vaporizes. The updraft created by the heat of the flame pulls the vapors upward, while

Lit by a single match, the flame of this piece of birch kindling is about to go out. Initially the flame burned higher because heat was concentrated in splinters and sharp irregularities on the surface. The heat caused gases to emerge freely. But now the flame is being defeated both by the increasingly larger mass into which the heat is dissipated and by the insulation afforded by the charcoal. A thinner piece of kindling—about finger thickness or less—would have been able to sustain its own flame.

oxygen from inrushing air mixes for combustion. The coincidence of vapors, oxygen, and 1000°F temperatures cause ignition. Yet the wick is not consumed, except gradually at its tip, because rising vapor wind prevents the inrushing air from reaching the wick. Thus, the right mixture for ignition occurs just on the edge of the cylinder-shaped stream of vapors around the wick, and only periodically touches the wick's tip, consuming it almost unnoticeably.

A burning piece of wood can function as its own paraffin and wick. As gases emerge from the wood, they ignite a short distance from the surface, because their own draft prevents combustion air from reaching the wood. As long as the flames from a single piece of wood can keep the wood itself hot enough to continue generating combustion gases, the flames will continue to rise.

But often a single piece of burning wood will not be able to sustain itself. This is a common problem at fire-starting time. Newspapers and small kindling start a crackling good fire. Yet by the time the kindling has consumed itself, the surface temperatures of a log will not have risen high enough to generate gases for self-sustaining flames. If the log does become hot enough to emit gases for flames, those flames may not be able to maintain surface temperatures necessary to generate gases. Thus, starved for combustible gases, the flame dies.

COMBUSTION AIR. A fire also needs sufficient combustion air. Or more precisely, the heated and fast-moving molecules of gases of the wood need to collide violently with the oxygen atoms that make up 21 percent of room and outdoor air. During these collisions new chemical bonds are formed that increase the number of types of combustible gases. Then combustion occurs only within specific limits for leanness or richness of the gas-air mix. For example, methane will ignite at about 1000°F only if it makes up from 5 to 15 percent of the mix with air. Hydrogen, on the other hand, is more cooperative; it can ignite at about 1000°F even if it makes up as little as 4 percent or as much as 75 percent of the gas-air mixture. At their own special ignition temperatures, most other combustible gases will burn when they make up anywhere from 10 to 60 percent of the mix.

Although a fire needs a supply of air, too much air at high velocity can kill the fire. This is most easily seen when you blow out a candle, forcing combustible gases away from the flame faster than the flame can catch up with them. This problem usually occurs in stoves and fireplaces only when starting a fire. Then flue drafts may be strong enough to put out matches. Or you can snuff out tinder for kindling by blowing too hard on it in an effort to provide combustion air. But once a fire has built up high temperatures in the wood itself, it would take a blast of air from an explosion to make the flames vanish. Here the blast would also be great enough to scatter the wood. Yet even with no visible flame to start it, most of this wood would still be hot enough to continue generating combustible gases. So wherever the wood pieces came to rest, the gases would emerge and mix with air at temperatures hot enough to rebuild a flame.

Air routing. For especially complete burns, a fire needs combustion air at two levels. First, it needs primary air entering either at the wood's own level or from beneath it. This primary air mixes with the gases as they emerge from the wood for initial ignition. Then there should be a source of secondary air entering above

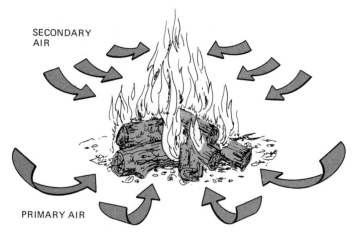

SECONDARY AIR

PRIMARY AIR

Fireplace fires usually result in fairly complete combustion because of the abundance of air sweeping from the room and up the flue. Primary air mixes where gases emerge from the logs, and secondary air allows ignition of hot unburned volatiles.

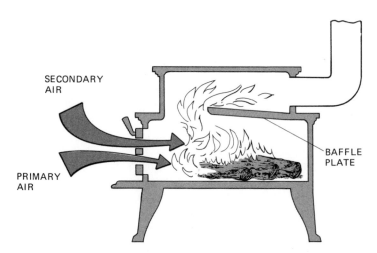

SECONDARY AIR

BAFFLE PLATE

PRIMARY AIR

Airtight wood stoves also allow fairly complete combustion when their air inlets are open wide. In this basic S-flow draft pattern, secondary air enters a turbulent chamber of flame and hot volatiles. This leads to ignition of most of the remaining volatiles.

the wood to mix with and to ignite unburned combustible gases, mists, and solid particles borne by the updraft. But unless the secondary air mixes before the unburned gases cool below their ignition temperatures, no flame will result. So the unburned gases, mists, and particles will travel up the flue. The more volatiles wasted up the flue, the lower the energy efficiency of the stove or fireplace.

Many fireplaces achieve fairly complete combustion because of the great volumes of air circulating around and through the fire. Stoves with air leaks also allow enough air in to maintain high rates of combustion, but the effectiveness of secondary air over the wood may depend largely on the locations of the air leaks. Airtight stoves and furnaces can be regulated to admit only limited amounts of combustion air, with a balancing of primary and secondary air. So, well-designed airtights can provide fairly complete combustion without robbing the house of as much warm air as fireplaces and leaky stoves do.

MOISTURE IN WOOD. A wood's moisture content has a pronounced effect on the way the wood burns. Any heat needed to boil off water is lost for ignition and home heating, and as long as a fire log is steaming and hissing heavily, the surfaces of the wood usually won't be hot enough to generate combustible gases. Then too, the steam itself dilutes other combustion gases and makes them less apt to ignite.

When freshly cut, some woods are more than half water by weight, and their cell structures may be so tight that they won't release the water until the wood has been split into thin pieces and given plenty of air circulation that results in evaporation. Other woods, especially those of dead standing trees, may have a relatively low moisture content, perhaps only 10 percent by weight. These drier, seasoned woods rob less heat from the existing fire for the boiling of water, and so more quickly heat up to produce combustible gases.

RESINS IN WOOD. Softwoods (conifers) are usually rich in tars and resins that vaporize and ignite at only 670°F—several hundred degrees lower than ignition temperatures of gases and other combustibles emitted by wood. Flames from the vaporized resins then can ignite gases of the hardwoods at their 1100°F ignition temperature. So it's easy to see why dry conifer wood and pine pitch itself are favored for starting campfires as well as indoor fires.

FLAME QUENCHERS. Cold surfaces can give a fire trouble. At fire-starting time, cold masonry walls in a fireplace and the cold metal of a stove keep nearby combustible gases below their ignition temperatures. In this phenomenon, called quenching, flames directed toward cold surfaces look as though they've hit an invisible shield just in front of the cold surface. This results from the sudden cooling of flame gases. The problem may be so severe that no flame may emerge at all from the side of a log near the cold surface. Quenching can be overcome, though, if the basic fire continues, for the radiated heat will gradually warm the cold surface enough so the flames can lap against it merrily.

Cold logs can also quench a fire, so it's advisable to warm wood indoors in winter, allowing it to reach room temperatures before placing it on the fire. The larger and colder the logs initially, the greater the problem.

USING WOOD THAT'S THE RIGHT SIZE. A matchstick will catch fire more readily than a larger piece of wood because the heat of its flame is concentrated in a small volume of wood. This raises the wood temperature high enough to quickly liberate combustible gases. But when the flame of a match is applied to the bottom of a large piece of wood, the heat is conducted into a relatively

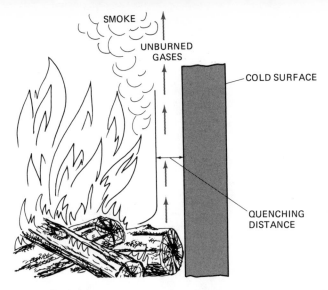

Cold surfaces near flames tend to quench the flames by keeping combustible gases below their ignition temperatures. This results in unburned particles and mists being borne upward as smoke, along with invisible unburned gases. This quenching effect is diminished and then eliminated as the surface is gradually warmed by the fire.

large amount of wood fiber. Here, the wood doesn't heat up enough to release combustible gases before the match has burned up.

Master firebuilders adjust their methods accordingly. In laying their kindling heap, they begin with loosely balled-up newspaper or with wood shavings. The wood fibers heat up quickly at the touch of a match flame. Then if flames can go to work on heating thin sticks, the sticks themselves will produce enough vapors to support flames. It's important that the kindling sticks be no more than finger-thick. This is so because a well-seasoned stick this size can sustain its own flame when lit from the bottom end. Then steadily larger pieces of wood can be added. In the case of a fireplace fire, it's wise to lay a back log before placing the kindling. Even though this back log may not catch fire itself for a long while, it will serve to protect the young fire from the quenching effect of back-wall masonry at first. Then it will reflect some heat to the fire, while gradually warming itself as the fire builds in intensity.

ALLOWING FOR ENOUGH AIR CIRCULATION. A stack of crumpled newspapers and beautiful kindling may fail to yield a decent fire if the newspapers are crumpled too tightly and the kindling is stacked too tightly to allow air to circulate between the sticks. Good starter fires can be made from kindling stacked like a teepee, or like a log cabin, or even in a haphazard mound. The important factors are size and dryness of kindling, spacing that allows air circulation, and one more factor: flame spacing.

Flame spacing. It's important to space wood pieces far enough apart to allow air to mix with vapors coming from the wood. But it's also important not to space the wood too much. The reason is that medium-size pieces and large logs often cannot sustain their own flames. They need the heat supplied by flames from

STAGES OF A WOOD FIRE

1. First, balled-up newspaper ignites loosely stacked kindling. Next, the small pieces of kindling will be heated up sufficiently, before the paper is consumed, to self-generate their own combustion gases and support their own flames.

2. The kindling shown in the previous photo is now ash and embers. The two pieces just above the embers were added to the kindling and allowed to "catch well" before the large piece on top was added. Flames from each piece of wood help heat the other pieces so that they continue generating volatiles.

3. Here most of the wood has been consumed, leaving mostly ash and charcoal. Later, a glowing charcoal fire will show mainly small, bluish flames, indicating a good air mix with carbon monoxide and small amounts of hydrogen. Eventually, only ash will remain.

nearby wood to keep their surface temperatures high enough for the generation of combustible gases.

So you should avoid one-log fires even though you've built up a hot bed of coals. Two logs will reflect and reinforce each other's heat. And three logs can do an even better job.

WHY CHARCOAL DOESN'T FLAME. After a wood fire has been going awhile, many of the pieces will show charred surfaces. These charred surfaces are charcoal. And as noted earlier in this chapter, charcoal is the result of burning in the absence of air. How can this be? After all, isn't the wood getting a steady supply of air?

No. The wood isn't getting the air because the gases from the heated wood are hitting the air, reacting with it, and igniting. So the charring wood is hot, all right, and it is undergoing pyrolysis (oxidizing without air), but it doesn't give off flames itself.

As the charred layer thickens, it serves to insulate the inner layers of wood from some of the intense heat. This slows the flow of gases from the inner layers and, ideally, results in an even, steady burn of the whole piece of wood. But sometimes this shell of charcoal retards enough heat transfer inward to stop the production of combustible gases. This is often the case with ends of logs in fireplace fires. Before they will catch flame again, they must be poked nearer the hottest part of the fire.

The reactions that drive gases from wood leave charcoal, which is composed mostly of carbon, and lesser amounts of hydrogen, oxygen, and some minerals that the original tree drank from the soil. Under continued heat, oxygen travels to the surface of the charcoal and there combines with carbon to form carbon dioxide, which doesn't flame, and carbon monoxide, which does. Some hydrogen may emerge too. The carbon monoxide and other minor amounts of gases ignite at temperatures over 1000°F. If flames occur, they are small, and they are generally bluish. With continued high temperatures and plenty of oxygen, charcoal gases continue to burn modestly until only ashes remain.

WHAT FLAME COLORS MEAN. Most types of wood can produce either blue flames or flames that vary between orange and yellow. When there's a good mix of oxygen with the other gases, the flame tends to be bluish. The characteristic blue flame from a gas-stove burner occurs because oxygen is premixed with the gas. The yellow-orange flames associated with young and medium-size wood fires result primarily from glowing carbon particles rising with the draft. These yellow-orange flames may be so bright that they give the illusion that the fire contains no bluish flames at all. But if you look closely, you can usually see blue flames too, especially where combustion air is mixing well with the hot gases coming from the wood.

Some people mistakenly regard yellow flames as sure evidence of incomplete combustion. But if the carbon particles causing the bright flame are being consumed there, as a result of a sufficient mix of gases with secondary air, they transform to carbon dioxide. A surer sign of incomplete burn is sooty smoke.

THE MESSAGES OF SMOKE. Smoke results when hot combustible gases, tars, and carbon particles aren't mixing well enough, or at high enough temperatures, with the air to ignite. This case is similar to that of a smoky candle flame. In a wood fire, smoke often occurs just above an area where the wood touches metal, masonry, or another log—again pointing up the importance of proper spacing for adequate air circulation.

Wood itself can cause smokiness. Generally, the resinous conifers are the worst offenders. But lightweight hardwoods such as aspen, cottonwood, and willow are often smoky too. For these woods especially, a good supply of secondary air can lessen the problem.

There is white smoke and black smoke. The white dense cloud that appears only momentarily over the chimney top and then vanishes is principally steam that is evaporating rapidly. True white smoke, though, and bluish smoke contain unburned tars and microscopic particles. The white smoke from a smoldering fire is essentially a tar mist. Darker smoke contains larger particles of fly ash and soot that will collect on the cold blade of a knife when you hold it over the flames.

Smoke is objectionable enough on grounds of air pollution. But it is also a signal of incomplete combustion and thus low heating efficiency from the burn. Lots of unburned gases, tars, and particles in the flue can also contribute to the buildup of creosote, the subject of the next chapter.

EFFECTS ON AIR QUALITY. Trees produce about the same amount of oxygen when growing as their wood consumes in rotting or burning. So wood burning doesn't really alter the atmosphere's oxygen level, other than to complete the cycle on an individual tree faster than rotting would. If forests are maintained for sustained yield, wood burning should have no long-term effect on world oxygen supplies.

Scientists fear that ever-increasing carbon-dioxide levels in the atmosphere may cause a global warming trend. Wood burning produces carbon dioxide, all right. But the amount is no greater than the carbon dioxide that trees absorb from the atmosphere in growing and then release in decaying. Here the planting of at least a few trees for each one burned can maintain a long-range balance of emission and absorption. The burning of fossil fuels, on the other hand, produces high amounts of carbon dioxide, and there is no counterbalancing absorption. The burning of grasslands and other agricultural burning has been a serious cause of carbon-dioxide emissions worldwide because there hasn't been enough replanting for a sustained yield of plants that absorb carbon dioxide. Still, if you burn wood for fuel, you are introducing the carbon dioxide into the atmosphere decades before that carbon dioxide would have entered as a result of the decaying of each tree. The effects of wood-burning emissions of carbon dioxide will, no doubt, be determined more precisely in coming years. For now, though, everyone can help with a counterbalancing tree-planting ethic.

Our unhealthy air. Until the mid-1970s, wood burning seemed destined for an increasingly smaller role in the home and industrial heating picture. Until then, money for research and sophisticated testing devices was devoted to fossil-fuel emissions, known to include dangerous sulfur oxides, nitrogen oxides, lead, and airborne particles. Prime offenders were automobiles and industry, with home heating being a lesser, though significant, source.

Then came the 1973 oil embargo and the world energy crisis. This naturally loosened U.S. governmental pursestrings for studies on alternatives to fossil fuels. Coal and wood again became plausible partial alternatives. Though emissions from coal were fairly well understood, not much was known about comparative emissions of wood and coal, aside from wood's reputation as an insignificant emitter of sulfur and nitrogen oxides, while being free of lead.

There's still great need for further research. But it's now clear that several wood-burning emissions deserve concern. These include carbon monoxide, airborne particles, and polycyclic organic materials (POMs). Let's look at each.

Carbon monoxide. This abundant poisonous gas results from incomplete combustion of all types of fuels. Carbon monoxide kills and sickens by combining with blood hemoglobin, in effect expelling oxygen. This deadly gas is an important automotive pollutant, and it has long been considered a hazard in small, airtight houses with low-burning overnight fires. Here lack of flue draft can allow carbon monoxide to reach toxic levels indoors.

Carbon monoxide ignites at temperatures of 1000°F and higher. Well-designed wood stoves and furnaces can be operated to consume far more of the carbon monoxide generated from wood than fireplaces and other wood-burning equipment can. Less-efficient wood burners can produce as much carbon monoxide as is emitted from well-designed coal burners, burning equivalent masses of bituminous coal. It follows that wood burned in the least effective equipment or in the wrong ways can significantly increase carbon-monoxide levels in communities already troubled with auto emissions. We'll cover ways of achieving relatively complete combustion of wood throughout this book.

Airborne particles. These are known to aggravate a wide range of respiratory problems and are among contributing causes of cancer. When burned in high-draft conditions, such as in fireplaces and during wood stove startups, wood tends to emit excessive amounts of particles in the form of visible smoke. Burned this way, wood can introduce more particles into the air than equivalent masses of coal, if the coal is burned more sensibly in well-designed coal burners, capable of relatively complete combustion at low draft settings.

Hot fire chambers with airflow patterns that mix airborne particles and unburned volatiles with combustion air can reduce particle emissions markedly. And there's another distant hope: Tests have shown that flue-filtering systems can trap nearly all particles. But so far these high-cost filters are available only for industrial use. We may someday see flue filters for homes.

Polycyclic organic materials (POMs). This grouping of hydrocarbons is suspected of being cancer-causing in combination with other common air pollutants. The small amount of testing done so far has shown highest emissions from the burning of resinous conifers, burned in fireplace conditions. Here emissions have been equal to those from equal masses of bituminous coal burned in well-designed coal-burners. These emissions have been sufficiently high to constitute an apparent health hazard.

More testing needs to be done before there will be reliable guidance for the reduction of wood-burning POM emissions.

Dispersion of pollutants. Like most substances harmful to health, the majority of air pollutants are not considered harmful in very low amounts. Researchers have suggested safe thresholds for both short-term and long-term exposure.

Time and extensive testing will help determine the validities of these thresholds.

Yet for many communities, the test of time may yield grim results. As an example, ski meccas in the Rockies have severe problems. In one ski town there are nearly 4000 fireplaces to serve the 4000 year-round inhabitants and an average 25,000 winter population. Here terrain features limit air movement, and thus the air becomes loaded with wood smoke and automobile emissions. Civic groups have financed air-quality studies in hopes of finding solutions—the most promising, though implausible, being the banning of automobiles and wood burning. So it is that ski towns and other small communities are grappling with air-quality problems formerly associated only with big cities and industrial towns. In some of these communities, wood burning can be a significant source of pollution.

All fuels pollute. Wind dispersion is a key to keeping pollutants at levels that are considered safe. As noted in the opening chapter, wood can be a more reasonable alternative for heating than fossil fuels. The key is to burn as little wood as possible to do the job.

3 | Creosote and Chimney Fires

ASKED FOR A quick definition of creosote, many people will describe it as a dark liquid used as a wood preservative. But most people who heat their homes with wood will offer a different off-the-cuff definition.

To wood burners, creosote is a dark, crusty substance that coats the insides of stoves, stovepipes, and chimney flues. Within the cleanout compartment at the base of a chimney, you can often find a small mound of fallen creosote fragments that look a lot like charred cornflakes and parts of prunes. Some of the fragments are shiny on one side. They give off an acrid smell. If you find chunks of creosote at the base of the flue, you can bet there's creosote lining the flue.

Creosote buildups can choke down the diameters of stovepipes and flues and so reduce draft. When creosote and ash line a stovepipe, they serve to insulate it, preventing desired heat transfer to the pipe and then to the room before hot flue gases exit up the chimney. But most important, if creosote is heated by direct flame or to temperatures over 1000°F, it can ignite and fuel a phenomenon that many people dread—the chimney fire.

But before we witness a chimney fire, let's take a closer look at the creosote itself. Its chemical makeup can vary, depending on the types of woods burned, conditions of the fire, and temperatures between the fire and the chimney top. Basically, creosote results as water vapor and airborne pyroligneous acids condense in a mixture that may have a watery or tarry consistency. Creosote contains acetic and formic acids, methanol, wood oils, and tars, among other compounds. Its high acid content makes it extremely corrosive to ordinary steel and mortar.

The hot flue gases of a wood fire carry water vapor and acids toward the chimney top. Along the way, if flue temperatures are cooler than about 150°F, the condensing creosote is watery. If flue temperatures are somewhat hotter, the water continues as vapor, condensing as creosote with a tarlike consistency. Very hot flues tend to collect less creosote because low temperatures causing condensation are not met until the gases have emerged out the flue top. But the creosote that does condense begins to bake dry and hard. If direct flames or high temperatures ignite the creosote, as occurs in a chimney fire, the creosote burns completely, leaving only carbon behind.

When a chimney fire occurs, flames in the flue create tremendous, thunderous updrafts that pull room air into the stove or fireplace with frightening power. Flames may blast from the chimney top, sending up a shower of flaming creosote fragments as well as molten mortar. These may fall on the roof and surrounding landscape.

If the chimney is of sound construction, the fire will eventually subside and

The first place to check for creosote is at the base of the flue. Accumulations of creosote like this are a sure sign of significant buildups higher up.

This is the underside of a flue cap as it was removed from the flue top. The cap's galvanized metal shows signs of corrosion after only one heating season.

finally go out. If neither the roof nor surrounding trees catch fire, the neighbors will go back indoors. The flue is now cleaner than it has been in a long time.

But the flue can be faulty: It may have cracks. Fallen mortar may have left spaces between joints in flue tiles. Metal flue liners may have corroded enough to let out flames and intense heat. And this last chimney fire, along with previous ones, may have finally caused enough deterioration of the flue to superheat adjacent parts of the house.

Many people start chimney fires deliberately by building hot fires or by tossing in compounds designed to remove soot and creosote by controlled burns. Some of these compounds have been known to explode in stoves. The intent is to keep accumulations of creosote fairly thin, so that the chimney fires are relatively small. These people look at the deliberate chimney fire as a kind of incendiary preventive maintenance. However, it's a questionable practice, because it sends fiery embers into the sky and gradually weakens the mortar. It can crack masonry, and it may heat up house timbers enough to start the house afire.

Firemen recommend that you try to keep your fires inside the stove or fireplace. You can periodically clean stovepipes and flues yourself or have it done by professional chimneysweeps. Techniques for this are described in Chapter 9. The maintenance of a chimney includes inspecting for defects as well as removing creosote.

There are special hazards with stovepipe fires. When stovepipe creosote ignites, it can turn the pipe red-hot, heating it to about 1400°F. This weakens the metal and shortens its life—which may not be more than a year anyway. But worse, a stovepipe fire can create enough of a gale inside to set the stovepipe dancing. Your problems at this point will be proportional to the length of the stovepipe, the amount of creosote to be burned, and the number of sheet-metal screws you forgot to install at pipe joints.

In the old days, people tried to achieve maximum heat transfer from the pipe to the room by using long stovepipes—up to 50 feet long. When some of those long pipes caught fire, they writhed like giant anacondas before breaking loose, spewing flame in all directions.

WHAT TO DO IF A CHIMNEY FIRE STARTS. First, notify the fire department, which would rather send a few men to monitor your routine chimney fire than send a whole company to put out a house fire that you tried to prevent by yourself. Until the firemen arrive, you can take a number of steps.

First, remind yourself and your family that the thunderous draft is a normal part of chimney fires. Remember that a flue in good condition can normally contain a chimney fire without letting the fire spread to the house.

Next, it's good to suffocate the fire in the stove or on the hearth. If you have an airtight stove, you can suffocate the stove fire simply by closing the air-inlet vents, starving the chimney fire of combustion air. If your stove is not airtight or if you have a fireplace, you can suffocate the fire by dumping sand or rock salt on it. Try to avoid tossing cold water onto the fire, though, for it will cause a messy splatter and the water may cool the metal or masonry quickly enough to crack it.

As long as there are any live embers in the stove or on the hearth, avoid closing off the damper leading to the chimney. Although closing the damper may help cut off combustion air in the chimney, it will also smoke up your house, maybe driving you outdoors, causing smoke damage, and reducing visibility for the firemen.

You can reduce the amount of combustion air entering a fireplace by blocking the opening with a wet blanket. This may take two persons, though, because the updraft may succeed in pulling the blanket loose.

Meanwhile, it's advisable to send someone outside to watch the flaming debris landing on the roof and on trees. If you can unroll a water hose, dampen the roof. Fortunately, well-insulated houses in areas of heavy winter snowfall will have a protective layer of snow on the roof.

The firemen will probably elect to let the chimney fire burn itself out if the chimney appears to be sound. If not, they may drop a smoke flare inside the flue to create an atmosphere that won't support flame. They'll also check the walls and roof adjacent to the chimney for evidence of overheating. If a wall prevents their getting a good look, they may ask your permission to punch through some wall paneling. Let them do it.

If all looks okay, the firemen will probably suggest that you keep a close watch on the areas near the chimney, at least for a few hours. If the incident unsettled you enough, you may elect to make periodic checks throughout the night.

No doubt you'll also vow to clean that chimney more often.

4 | Home Heating Principles: Airflow, Space, and Insulation

FIRE IN A STOVE or fireplace heats a home by means of three main forms of heat transfer: conduction, convection, and radiation. Then this heat is retained in the home mainly by materials and design features that retard heat transfer to the outdoors. So your challenge becomes one of promoting maximum heat transfer around the fire and then discouraging heat transfer through the house exterior.

Before we take a look at kinds of heat transfer, let's make a distinction between *temperature* and *heat*. Temperature indicates the speed at which molecules are in motion. The faster these molecules move, the higher their temperature in that location. All gases, liquids, and solids are made up of molecules in motion. The amount of motion (which is the same as saying the temperature) determines whether a substance takes on a solid, liquid, or gaseous form.

Water is a familiar example. Below 32°F, water is solid ice. Its molecules do remain in motion even though their temperature may be lowered hundreds of degrees further, but they slow more and more as the temperature drops. Finally all motions stops at −460°F, absolute zero. At 32°F and up to 212°F, water has a liquid state. Above 212°F the water becomes a gas; its molecules are moving at high speed and occupy far more space than at lower temperatures. This makes the molecules lighter in weight for a given amount of space—making them more buoyant in normal room air.

Heat, though, depends on the *amount* of molecules as well as the intensity of their motion. For example, a piece of aluminum foil can have a very high temperature. Yet it may hold only a little heat for a short while because it is so thin. So a good heater needs high temperature, and it needs enough mass to allow heat storage, which usually improves steadiness of heat.

HEAT TRANSFER. Heat always travels toward cold, and this is accomplished by three means of interest to people who burn wood.

Conduction. When molecules are heated, they begin to collide more vigorously with nearby molecules. A hotter molecule tends to share its heat with the cooler molecules it bumps, so that after the collision they have nearly equal temperatures—the givers a little cooler than before, the receivers a little warmer.

When a fire's heat is applied to a metal stove wall, the molecules of steel or cast iron become highly active near the source of heat. Their collisions set up a radiating chain of collisions even though the molecules themselves remain about where they were to start with. The temperature at the corner of the stove farthest from the heat source becomes nearly as high as that closest to the source.

So when heat is transferred through solids, liquids, and gases while the molecules remain pretty much in the same place, we're talking about conduction.

Convection. Here heated molecules do the traveling. This is easily seen in liquids and gases that are conveyed by circulating warm and cold currents. Heated air rises until stopped by the ceiling or until its warm air molecules have dissipated into cooler surrounding air. The flow of warm buoyant air and the descent of cool heavy air are called *natural convection,* or sometimes *gravity convection.*

Radiation. There is no such thing as radiant heat. But there is radiant energy. This energy travels by infrared electromagnetic waves at the speed of light. It doesn't heat the air it travels through; it heats only the surfaces it strikes.

In conduction, the molecules collide in chain reactions but remain essentially in the same place. Example: Heat is transferred from the stove wall nearest the fire to the farthest corners of the stove wall.

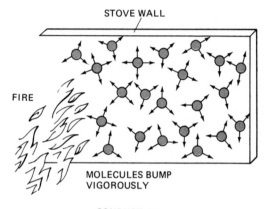

CONDUCTION

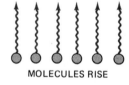

MOLECULES RISE

CONVECTION

In convection, the molecules themselves become buoyant and travel upward. Example: Air warmed near a stove rises.

In radiation, infrared electromagnetic waves from the fire or stove travel through air, heating objects they strike without directly heating the air. Example: When you hold your palm near a flame, your palm may feel warm while the back of your hand still feels cool.

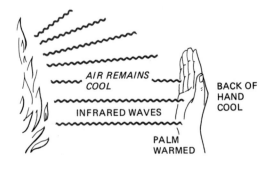

RADIATION

You've no doubt felt heat of radiant energy when sitting near a stove or fire. The nearer you sit, the more radiated energy strikes your skin. If you hold the palms of your hands toward the heat source, the skin of your palms absorbs the energy and you feel it as heat. Yet the backs of your hands may still feel cool, or even chilly. This is so because the air itself has not been heated.

Fireplaces and stoves with single shells tend to heat primarily by radiation. This radiant energy warms room air for the most part by conduction *after the energy has heated the surfaces of ceiling, walls, floor, and other masses in the room.* This air, heated by *conduction* of heat from various surfaces, becomes buoyant, rising by *convection* to other parts of the house. This convected air then transfers heat by *conduction* to cooler air as the warm air molecules collide with cooler ones.

VARIATIONS IN HOUSE DESIGN AND CONSTRUCTION. Once heat is introduced into a house, several factors will determine whether the temperature is comfortable.

House shape. When insulation, weatherproofing, and floor space are equal, a dome-shaped house is easier to heat than a boxy house because the dome offers less area for heat loss through its outer shell. Less surface area means less chance for heat to escape by infiltration and simple conduction right through insulation.

Of the basically box-shaped houses, a simple box has less surface area than a rambling house of the same floor space, and a two-story box-shaped house has less exposed surface than a one-story box with the same floor space. A two-story house has the added advantage of potentially better flow of convection air.

House size. Generally, the greater the volume of a house the more fuel will be needed for room comfort. This is so partly because there's more air to heat and partly because of greater heat loss through the larger exterior surface. In the process, rooms farthest from the heat source tend to be cooler than rooms closest. Here heat transfer can never catch up with the amount of heat escaping through the house shell.

If you own a large house, you can greatly reduce the heating load by being satisfied with room temperatures of 68°F or even 65°F, rather than temperatures over 70°F. Additional savings result when you close off unused rooms. Bedrooms, particularly, can be closed off during the day and allowed to warm up a bit just before bedtime. In fact, in Europe, many families customarily sleep "cold." That is, they seldom let any warm air into the bedrooms. Hot-water bottles prepare beds for occupancy, and down comforters help retain body heat. Sleeping this way, though, you may want to wear a nightcap as well as taking one before you retire. Sleeping cold requires some getting used to, but it sure reduces the heating load.

Then again, if you find you are living in a big house with more than a few excess rooms, there may be long-term economies in selling the big home and buying a smaller one. Or if you live in a large older home in need of roofing or exterior remodeling anyway, you may be able to "saw off" the top story or back rooms, reducing the amount of exterior materials needed to refinish the house. Resultant savings on exterior materials may nearly offset the contractor's fees for the changes. And if you do the work yourself, you may be many dollars ahead, even before the fuel savings during the following winter.

Circulation patterns. We'll be examining this subject in more detail in later chapters, but for now let's take a look at some basic principles of air circulation that affect heat distribution in most homes.

Warm air tends to rise, creating increased pressures in upper stories. This increased pressure, together with warm air's determination to mix with cold air, makes the warm air exit through cracks in upper stories. The result is lowered pressures at lower levels, and this draws in cold outdoor air.

The tighter the house, the longer warm air lingers at upper levels. With less warm air exiting at upper levels, less cold air infiltrates at lower levels. Happily, the result is a less drafty house and lessened demands on the heat source for living comfort.

In a tight, well-insulated house, you can take advantage of natural convection currents. In this case, heated air at upper levels cools gradually near ceilings and walls, rather than quickly escaping through cracks. As this warm air cools, it is replaced by a steady supply of warm air rising from the heat source. Cooling air descends by the path of least resistance and is forced near enough to the heat source to be of use in two ways: It may be drawn into the fire as combustion air, and it may be reheated and recycled through natural convection patterns.

You can encourage this sort of recycling for more even heat distribution by proper placement of heating ducts. An open stairway can serve as an excellent

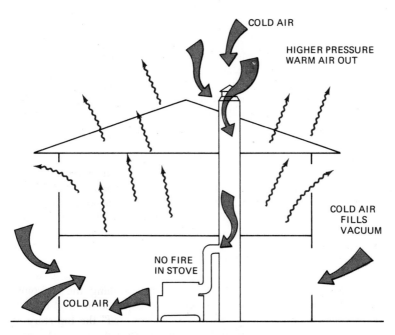

Rising warm air creates higher pressure upstairs, leaving lower pressure downstairs. Outside air then rushes into the low-pressure area, causing cold drafs downstairs. Outside air may even enter through a cold chimney and stove. These circulation patterns evidence the "stack effect."

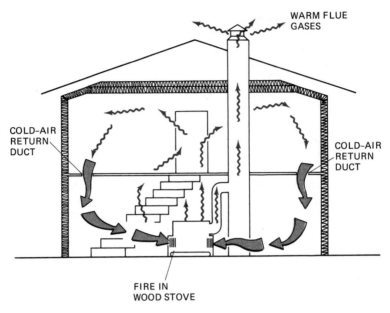

WARM FLUE GASES

COLD-AIR RETURN DUCT

COLD-AIR RETURN DUCT

FIRE IN WOOD STOVE

In a tight, well-insulated house, warm air near the ceiling cools and descends. This same air will recirculate near the stove if house shape and ducting encourage the phenomenon.

duct for rising warm air. Then you can install cool-air grates at appropriate distances from the stairwell to channel cooling air and promote convection recycling. This will minimize dead warm-air pockets upstairs and dead cold-air pockets downstairs. Or you may decide to install a hood right over the heat source topped by an air grate leading upstairs. Then cold-air return grates can be placed appropriate distances away.

If your central heating system employs warm air and ducts, you can use the furnace blower to force air through the duct network. Each hot-air register can be damped to balance the system the way you want it. Of course, the blower consumes electrical power. A 500-watt blower would consume 10 kilowatt-hours during a twenty-hour day—costing you $15 a month. But you need not keep the blower going around the clock. It can be switched on to stimulate desired heat distribution.

INSULATION AND WEATHERTIGHTNESS. No matter what kind of house you have or what type (or combination of types) of heating system you use, you can increase effectiveness of the system by beefing up insulation to recommended standards, and you can further increase effectiveness by minimizing air leaks in the shell of your house.

If you already have a reasonably reliable heating system, plus a backup system, your thoughts should turn to insulating and weatherstripping. But if you have only one means of heating, and if you have a limited budget, you may have to choose between insulating and a wood heater.

Today, with the real possibilities of fossil-fuel shortages, it may be wise to opt for the wood heater first, and vow to beef up insulation later, as you can afford it. Besides, a wood fire transforms a house into a new kind of home. You could never gain as much enjoyment on a cold winter's evening sitting near the insulation.

In many ways, home insulation functions like the fill in a sleeping bag. Its capacity to keep heat inside is determined by the thickness of the dead-air space between the interior and the exterior. But some dead-air spaces are deader than others. If the dead air is enclosed in cells that prevent any air movement and transfer of heat through the air by convection, the insulation can be highly effective per inch of thickness.

Before a sleeping bag can perform up to its insulating potential, it must be fluffed up so that the fill provides maximum "loft"—in other words, maximum thickness of dead-air space. But when slept in, that sleeping bag can fail miserably on its underside because a person's weight compresses the fill, greatly reducing the space between person and frozen ground. The most popular way to compensate for this deficiency is to place a urethane sleeping pad underneath the bag. This pad contains tiny air bubbles sealed in the synthetic. Body weight scarcely compresses the pad, leaving enough dead-air space to reduce heat loss from the person to the ground.

There are home-insulation materials that perform much like the fills in sleeping bags and the closed cells in sleeping pads. These home-insulating materials are assigned R values that correspond to their resistance to heat flow. For example, a ½-inch sheet of plywood siding has an R value of less than 1, meaning that heat travels through it rather easily. An 8-inch concrete block with hollow cores and about 4 inches of aggregate resists heat transfer only a little better than ½ inch of plywood; the R value of the block is only 1.11. As the accompanying table indicates, just 3½ to 4 inches of glass fiber insulation has an R value of 11, and it takes only 1½ inches of rigid urethane foam for the same R value.

These R values are additive. That is, for every 3½ to 4 inches of glass fiber, you can figure on an additional R-11 value. So, if you insulated with 12 to 13 inches of glass fiber, you could figure on more than tripling that R-11 value. For that 12 to 13 inches of thickness, glass fiber is awarded a rating of R-38. Yet it takes only about 5½ inches of rigid urethane to achieve an R-38 value.

R VALUES OF INSULATING MATERIALS

	BATTS OR BLANKETS		LOOSE FILL (POURED IN)			RIGID PLASTIC FOAM		
	GLASS FIBER	ROCK WOOL	GLASS FIBER	ROCK WOOL	CELLU-LOSIC FIBER	URE-THANE	U-F	STY-RENE
R-11	3½"-4"	3"	5"	4"	3"	1½"	2"	2¼"
R-19	6"-6½"	5¼"	8"-9"	6"-7"	5"	2¾"	3¾"	4¼"
R-22	6½"	6"	10"	7"-8"	6"	3"	4"	4½"
R-30	9½"-10½"	9"	13"-14"	10"-11"	8"	4½"	5¾"	6½"
R-38	12"-13"	10½"	17"-18"	13"-14"	10"-11"	5½"	7½"	8½"

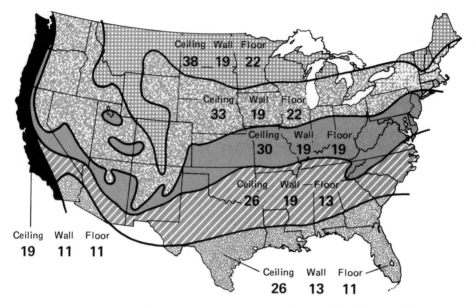

Ceiling Wall Floor
38   19   22

Ceiling Wall Floor
33   19   22

Ceiling Wall Floor
30   19   19

Ceiling Wall Floor
26   19   13

Ceiling Wall Floor
19   11   11

Ceiling Wall Floor
26   13   11

This map shows R values recommended by Owens-Corning for highest net savings on both heating and cooling costs.

As the accompanying map indicates, R values of R-38 are recommended for attics in homes in the coldest northern states. Farther south, lower R values are deemed sufficient. The reason attics need more insulation is that warm air is buoyant. It floats upward through registers, through floors, up stairwells. Momentarily halted by the attic insulation, this warm air begins to escape through leaks in the ceilings and walls, and heat is transmitted by conduction right through the insulation itself.

Can't insulation and weather seals prevent heat loss? No. They can retard it greatly, but not prevent it. In a tight and well-insulated house, about 50 percent of the heat is transmitted outdoors by conduction—that is, by the warming of materials in the house shell and consequent transmission of heat through their outer surfaces. The other 50 percent simply leaks out through cracks—and for every cubic foot of air that leaks out, a cubic foot of cold outside air replaces it.

In a drafty house, you can expect about two complete air exchanges per hour. A tight house with good insulation and proper thermal windows will have about one complete air change every hour. In a superinsulated house, carefully designed to conserve energy, about the best you can hope for is half an air change per hour. Yet you need almost that much if you want fresh breathable air.

Insulating an attic. Studies have shown that properly laid attic insulation will pay for itself in fuel savings during the first heating season. Since unfinished attics are usually accessible, they often pose the easiest insulating task for the do-it-yourselfer. Finished attics pose additional challenges, and we'll allude to them shortly in the section on walls.

For unfinished attics, popular choices include batts and blankets of mineral or glass fiber. Also popular are loose-fill particles that can be poured between ceiling joists. But before you lay or pour any insulation, consider the vapor barrier.

Vapor barrier. This is a layer of impermeability. It may be paper, aluminum foil, or even enamel paint. Its purpose is to prevent moisture from escaping, which results in two main benefits. First, it prevents moisture from condensing on structural members and sheathing. In winter, such condensation can cause heavy frost buildup and ice damage. Then, come spring, that same frozen condensate melts, soaking into the woodwork and eventually causing rot.

The second benefit is the retention of water vapor in room air. Studies have shown that humid air feels warmer than dry air of the same temperature. This is due to the evaporative cooling that dry air causes on a person's skin. Thus the higher humidity retained by the vapor barrier can result in fuel savings because people feel just about as comfortable in a humid house as they do in a drier house with air a couple of degrees warmer. *Note:* Don't confuse the fuel economies of retaining humidity with humidifying itself. It's one thing to let the vapor barrier passively retain what humidity there is. It's quite another thing to use an electric humidifier. These humidifiers consume energy, and the resultant vapors absorb room heat as they warm to room temperature. The misty air near a humidifier actually feels cold on the skin because it is drawing heat from your skin. About the best way to humidify is by keeping a tea kettle on the stove.

Some insulation materials, such as urethane and styrene, are impermeable of themselves, so they don't require an additional vapor barrier. Batt and blanket insulation can be purchased with either a paper or an aluminum-foil vapor barrier on one side. Batts and blankets with a vapor barrier on one side are called *faced insulations.* Batts and blankets without a vapor barrier are *unfaced.*

If you wish to add insulation to an inadequately insulated attic or wall, check first to see if the old insulation has an effective vapor barrier. If it does, you

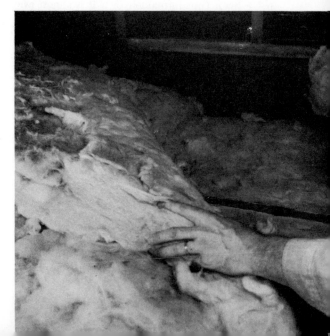

In this attic, two 6-inch layers of mineral wool provide an R-38 insulating value. The first layer, between joists, is faced with a foil vapor barrier on the warm side. The second layer is unfaced and is laid perpendicular to the first to minimize air leaks between rows.

can just add insulation on the cold side without any vapor barrier. But if the old insulation has no vapor barrier, it's best to remove the old stuff and put down a good vapor seal. In extremely cold northern climates, a two-layer, overlapping vapor barrier is recommended. Remember, always keep the vapor barrier on the warm side of the insulation.

Insulating walls. If interior walls are ice-cold in the winter, they probably lack insulation. The old solution was simply to tear out wall panels, install blanket insulation, and cover with new wall panels. That may still be your only option if horizontal fire stops have been nailed between wall studs.

But if there are no fire stops to prevent it, you can bore holes near the ceiling between wall studs and pour in loose-fill insulation. Since the house will be heated with wood, it's smart to avoid flammable loose fills made of wood fibers. Use rock wool or glass particles instead. When poured from the top of the wall, the loose fill trickles down around electrical wiring and boxes to form chambers of dead air as thick as your wall studs. Since 2×4 studs really are only 3½ inches deep, loose fill there can give you only about R-8. But 2×6s (5½ inches deep) can get you up to about R-11, still below the recommended R-19 value for walls that can be cost-effective for most parts of North America.

So noncombustible loose fill alone can't provide optimum insulation for northern homes with 2×4 studs or even with 2×6s. This leaves the possibility of adding rigid foam panels to the inside walls. The trouble is, an added thickness of interior paneling usually requires major modifications in moldings and trim in order to retain an attractive interior. Rigid urethane is the best insulator per inch (just 1½ inches give R-11 value). There are flame-resistant and nontreated types. Both give off toxic fumes once they begin to burn. Styrene, the other rigid-panel option, is also available in flame-resistant and nontreated forms; it too will burn, though, and emit toxic fumes. The Federal Trade Commission recommends that all urethane and styrene be covered with ½-inch flame-and-heat-resistant gypsum board.

Perhaps the most controversial, and most expensive, insulation for finished walls is UF foam (ureaformaldehyde). It is almost as effective an insulation as urethane per inch (2 inches for R-11, and 3¾ inches for R-19). This is the liquid foam shot from spray nozzles, normally from outside the house. As a bonus, the foam has proved fire-retardant, though it's not "fireproof." *But caution:* Mixed incorrectly, this stuff has been known to give off a lingering unpleasant odor, and it can shrink after setting up. Be sure you receive a written contract that gives you recourse in the event of either misfortune.

Insulating a crawl space. A good choice for insulation here is batt and blanket. This is so because of low cost and ease of installation, and the fact that there's no concern for appearance.

There are essentially two ways to go. You can insulate the crawl space itself or you can blanket the underside of the first floor, leaving the crawl space cold. If you choose to insulate the crawl space, ensure that there's a good vapor barrier of polyethylene plastic on the dirt floor. The plastic should have generous overlaps that are weighted down with planking or bricks. The wall should be covered with faced batts or blankets, with the vapor barrier facing the warm side. The insulation should then extend over the floor a couple of feet.

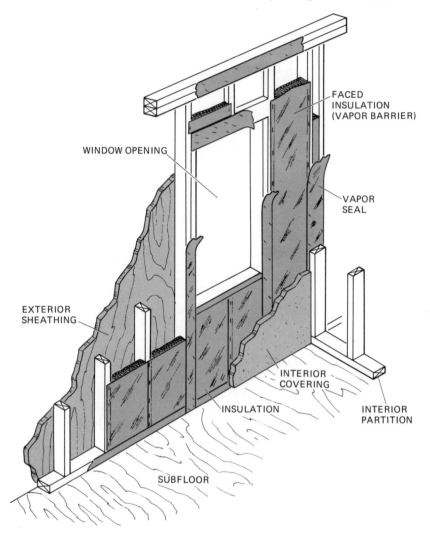

Walls in new houses are often insulated with mineral-wool blankets. Note that vapor seals are applied to wall framing to overlap the vapor barrier that comes attached to the insulating fibers themselves. In existing homes, old wall panels must be removed to allow this type of insulation to be installed.

If you choose to insulate the underside of the first floor, stuff faced insulation between floor joists, face up. Allow a small air space between the floorboards and the insulation. Then, since the flanges designed for stapling will be hidden on the

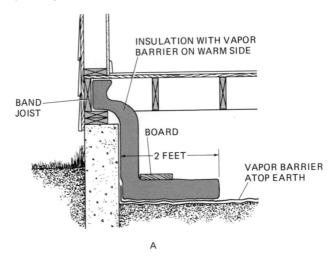

A

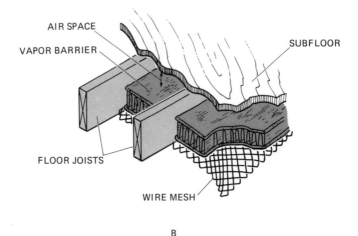

B

Crawl spaces may be insulated themselves (A), or you may insulate only the underside of the floor (B).

topside of the insulation, you can secure the blankets by stapling chicken wire along the underside of the floor joists.

Insulating basement walls. Concrete blocks and poured concrete are lousy insulators. As noted earlier, hollow-core blocks 8 inches thick provide an insulating value of only about R-1. Poured concrete 8 inches thick does worse. Neither material 8 inches thick has the insulating value of a mere inch of softwood, which is a little over R-1.

Many well-meaning people insulate basement walls, with disastrous results. It's true that floor-to-ceiling insulation in a below-ground basement makes the basement cozier, but it also prevents sufficient heat from escaping into the surrounding earth. Without benefit of thawing warmth from the basement walls, the adjoining soil can freeze. Especially in extremely cold weather, this frozen ground can develop tremendous force from ice expansion. The result is often cracked walls, or worse, caved-in walls.

Best advice in cold northern regions is to insulate only to about 2 feet below ground level. Of course, if an entire basement wall is above ground level, then you can insulate right to the floor. Again, if your wood heater is in the basement, it is safest to insulate with noncombustible materials.

It may be questionable to install a fireplace or a radiant-type stove in the basement, though. Many people do this, figuring that hot air rises. But radiant-type stoves and fireplaces heat primarily by means of radiant energy—emitted in straight rays. Much of the radiant energy travels straight to basement walls, and much of the heat generated on the wall surface is drawn quickly through the outer surface.

Circulating-type stoves tend to perform more efficiently in basements because the jacket surrounding the firebox absorbs the radiant energy. This heats up the jacket, which releases heat by conduction to circulating air. The heated air becomes buoyant and rises by convection toward the basement ceiling. Here a hot-air register or a stairwell can route the warm air upstairs. Wood-burning boilers and furnaces also work well in basements, because nearly all heat is transferred to hot-water tanks or routed through ducts.

All-weather windows. On the average, well-sealed double-glazed windows will lose about six to ten times as much heat as an insulated wall section of equivalent area. So energy-conscious people are examining ways of eliminating and minimizing the sizes of the most energy-wasteful windows in their houses.

Though you should consider the chill factor of the prevailing wind, the key concern is the sun. On sunny winter days, a single-pane window on the south side of a house will gain about as much solar energy for the house as it loses overnight. If that same window is double-glazed, it will gain about twice as much energy as it loses overnight. So if the sun shines nearly every day in winter, a double-glazed southerly window will gain heat. If the sun shines only half the time, the same window will be about as effective as superinsulation.

You need windows for natural light and for a view of the outdoors. Yet if you can reduce the size or number of northerly windows, you'll achieve energy savings.

Caulking and weatherstripping. As warm air escapes the house, it leaves a momentary pressure void that is quickly filled by outdoor air entering through any available openings. The warm buoyant air exits mainly through upper portions of the house. Cold air enters wherever air pressure is lowest, usually through cracks and porous materials in lower portions of the house.

By tightening the upper portion of your house first, you reduce the dynamics of the warm-air-exit, cold-air-replacement phenomenon called infiltration. Reduced infiltration results in warmer air and less draft.

You can recover the cost of materials for tightening up your house in a single year. In some homes, the fuel savings may be as much as 25 percent.

Weatherstripping. This involves sealing the gaps between moving parts of doors and windows. Possible materials include rolled vinyl, spring steel, felt ribbons, or strips of foam that can be tacked or taped in place. Vinyl and steel are preferred for sliding windows because they don't abrade the way felt and foam do. All four types can be used for a compression seal, as when a door or window closes against the weatherstripping. Price is often commensurate with effectiveness and durability. You'll have to weigh your budget against factors such as appearance, ease of installation, and effectiveness. Remember, any weatherstripping correctly installed will become cost-effective in less than a year. This includes door thresholds.

Caution: As a wood burner, you should avoid sealing up your house too tightly. A wood fire needs fresh combustion air and continuously draws room air up the flue. Thus often it's wise to neglect weatherstripping near the fireplace or stove. This way fresh air runs a short path to the fire and doesn't become a long, cold draft running through one or more rooms. Then too, a very tight house can result in reduced draft in the flue, enough to increase the risk of carbon monoxide backing up from a low-burning charcoal fire.

Caulking. This is the gooey substance designed to seal nonmoving gaps up to about ⅜ inch wide. It is commonly sold in cardboard cylinders that load into an inexpensive applicator gun. The caulking is forced out a nozzle, much the way decorative cake frosting is.

Beware of inexpensive oil-based caulks selling for about $1 a tube. They are popular because they are cheap. But these caulks dry out within a couple of years, crack, and lose their effectiveness. Silicone-rubber caulks cost more but are far longer-lived. Be sure to check labels to ensure that you can paint over the caulk.

The most expensive caulk is urethane. It applies easily and can fill large cracks, and you can paint it. It even holds fast in cracks subject to some movement. Consult the labels before you buy.

For gaps over ½ inch, you can first jam in oakum rope. It's the same rope used for chinking logs in modern cabins.

Caulk should be applied to prevent air and water leaks. In this photo, a bead of caulk is being drawn along a crack between a concrete-block chimney and a wall. Without the caulk, water can enter and eventually cause ice damage.

5 | Fireplaces

FEW PEOPLE CAN resist the allure of a fireplace fire. Since man domesticated fire some 750,000 years ago, open fires have cooked food, provided warmth, entertained the eyes, encouraged deep thoughts, drawn people together, promoted soft talk, and generally made people glad to be near.

Frank Rowsome captures the essence of an open fire in his often poetical book *The Bright and Glowing Place*. He pronounces a fireplace "inherently hospitable," a place to "renew an old interrupted friendship, or as a practical if delicate matter, to see if you want to renew it." Rowsome suggests that a fire gives "a deep sense of belonging," something more than "mere moving images after dinner," rather a symbol of the "core of the concept called home."

Indeed, home-decorating publications rarely miss a chance to toss in a working fireplace to capture our hearts and serve notice that the designers appreciate the best things too. Who could be indifferent to the messages of the hearth? What winter's evening is not improved by the crackle of a fire and a hint of wood smoke? Once the fire has toasted your palms and your backside beyond comfort, you can turn to give your frontside an equal share. All this while reflections of the flames dance on you, and—if it's right—someone you love.

THE TROUBLE WITH FIREPLACES. Even though they meet many human needs, fireplaces of traditional designs can be terrible heat wasters. In this age of energy transition, the fireplace is "under fire." Yet its long-hallowed image leaves us guilt-ridden as we reluctantly concede its shortcomings.

The popular press often attributes efficiencies of 0 to 10 percent to fireplaces. The usual explanation for this figure is often simple, but misleading. Usually, the explanation is that 0 to 10 percent of the potential energy of the wood heats the house and the rest goes up the flue.

In truth, fireplaces can be designed to deliver more than 10 percent of the wood's potential heat. This, combined with the 5 to 15 percent of the wood's heat that can be conducted through the walls of a chimney rising through the center of the house, gives many people all the ammunition they want to justify heating by means of a fireplace.

But there's a catch, and it's a big one: For proper draft, open fireplaces draw tremendous amounts of warm room air up the flue. The Federal Energy Administration warns that open fireplaces can increase home heating bills. Here's why:

On a cold day with the fireplace going and the furnace firing, a fireplace often will draw about 200 cubic feet of warm room air up the flue every minute. Some draw less than 200 and some draw up to 300 cubic feet. This causes cold outside air to infiltrate, replacing the lost warm air. The thermostat senses the temperature drop and keeps the furnace firing. The colder the infiltrating air, the greater the load on the furnace. Of course, the furnace may not fire as long if the thermostat is set at 50°F rather than 65°F. Or if the thermostat is placed in the room with the fireplace, it will sense a warmer temperature than it would in other rooms, and so it won't fire the furnace as much. Yet the fireplace will still rob whatever volume of warm room air it needs for draft.

FIREPLACES AND AIR EXCHANGE. In the previous chapter, we noted that a fairly well-insulated and weathersealed house will undergo a complete exchange with outdoors air about every hour. A superinsulated and supertight house can hold this exchange to a two-hour cycle. Again, this results from warm room air leaking out and cold outdoor air replacing it.

Let's say you live in a well-sealed one-story house measuring 30×40 feet with an 8-foot ceiling. This house would contain about 9600 cubic feet of living space. Without a fireplace, such a house would undergo a complete air exchange every hour. With a fireplace, drawing air at a rate of 200 cubic feet per minute, the frequency of complete air exchanges would more than double.

No wonder the medieval castles were drafty. No wonder the British later regarded working fireplaces as excellent means of drawing off vapors of impure, once-breathed air thought to cause all manner of diseases. But today, a 20 percent air-exchange rate per hour is considered sufficient for fresh air and good health.

How to minimize air exchange. One popular answer is to fit a screen of tempered glass over the fireplace opening. Then air sufficient for combustion can be admitted through louvers at the base of the screen. This keeps the fire going, while greatly reducing the loss of room air. The trouble is, the glass doesn't allow radiant energy to get back into the room. The glass absorbs some

Louvered glass screens can greatly reduce the amount of warm room air lost up the flue. This unit has been fitted with the Aglow Heat-X-Changer, whose intake (bottom left) and exhaust (bottom right) force warm room air through heat-exchange tubes under the fire and then back to the room as warmer air.

of the energy, reflects much of it, and basically results in a fire that can be seen but not felt except through the wall of the chimney if it rises inside the house. The wood burns up, and that's that.

Without a tempered-glass screen, you can reduce the loss of room air by means of a metal damper in the fireplace throat. Most dampers have a hinged door or pivoting flap that can be opened to various settings by means of levers, pull chains, or screw-type controls. Of course, the greater the opening, the greater the draft, and increased draft results in more loss of room air. So to reduce the air loss, you should close the damper to the smallest opening that still carries off the smoke.

The U.S. government estimates there are 25 million fireplaces in the country. Of these, the percentage without dampers is not known. But I do know that

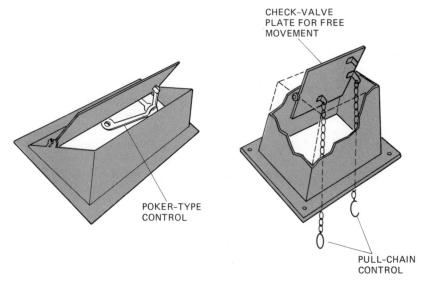

CHECK-VALVE
PLATE FOR FREE
MOVEMENT

POKER-TYPE
CONTROL

PULL-CHAIN
CONTROL

Here are examples of the many dampers available for fireplace throats. A damper allows control of draft while the fire is going, and it can prevent the loss of room air up the flue when the fire is out.

fellows who custom-make dampers for existing fireplaces do a thumping good business before every heating season. Many of these dampers are going into old fireplaces that have never had a damper.

The damper can save fuel in several ways. First, as we've seen, it can reduce the amount of warm room air escaping while the fire is going. This retards much of the infiltration of outside air that would otherwise occur and cause the thermostat to fire the furnace continuously. Second, the damper can reduce the rate of wood consumption. Third, and perhaps most important, it can stop the escape of warm room air when there's no fire.

Substitutes for a damper when there's no fire include wads of newspaper or commercial insulation stuffed into the fireplace throat. This can work pretty well until it's time to light the fire—and many a fire has been built with the throat still plugged with insulation. If accompanied by tall flames, such fire may prevent the removal of the insulation. Or it may simply smoke up the house—mysteriously. This can be quite humorous, if it's in someone else's house.

A well-designed fireplace can contribute noticeable amounts of heat to the home. But it can result in heavy losses of room heat overnight. Let's say the family has enjoyed a pleasant fire during the evening. Come bedtime, the choices are to let the fire die or to bank it, using embers, a mix of seasoned and green woods, and an insulating layer of ash to ensure a slow, even burn. If you let the fire die by scattering the remaining wood, you'll have some wood left for the next fire. If you bank the fire expertly, you'll have a hot bed of embers in the morning that will need only a few fresh sticks of wood and maybe a poke for a quick morning fire. That early-morning blaze with a cup of coffee has graced the pages of many an evocative book.

Trouble is, a banked fire will lose far more room air up the flue than it contributes to the house. And if you choose to let an evening fire die on its own, the open damper will release warm room air throughout the night. Yet, while the fire remains, you can't close the damper completely without smoking up the house.

With damper but no glass screen. To stop overnight heat loss, you should ensure that the fire is out before you retire. If only a few embers remain, you should extinguish them. Charcoal gives off toxic carbon monoxide. So with little to no flue draft there is some danger of fumes, especially if your house is small and tight, allowing minimal air exchange with outdoor air. Some fireplaces have ash dumps in their bottoms, but hot ash is probably best burned on the hearth. Then too, you can always shovel hot embers into a lidded scuttle and carry them outside for the night. But, by all means, find a safe way to close the damper for the night.

With a glass screen. This is better than no screen. At bedtime you can simply close the screen and close off the air-inlet louvers too. The fire may continue for a while on flue air or on air leaks in the screen louvers. But this way there's little loss of room air up the flue. The chimney may continue to radiate heat into the house. The only loss is that of wood essentially wasted as it burns up. If the fire expires before the wood has burned completely, you'll have wood for use in the morning.

FIREPLACE DESIGN. You've probably seen the extremes in possible designs yourself. There's the fireplace that seems to be a low, shadowy void with andirons placed far back, almost at the back of a cave. There's the tall, shallow Rumford with the andirons practically jutting into the room. In between are compromises of sorts that comply with Department of Agriculture guidelines or with dimensions used in prefabricated metal fireplaces. All have their advocates. But as far as I can determine, no researcher or agency has published results of tests that demonstrate the superiority of one design over the others.

The low, deep tunnel design is a favorite of masons who don't want their names associated with smoky fireplaces. Most of the radiant energy in these fireplaces is absorbed by the restrictive walls and the low lintel. Having absorbed the energy, the heated masonry allows incoming air to conduct the heat and carry it up the flue. Although these deep-cave fireplaces emit very little radiant energy, they often require only modest amounts of warm room air to keep the fire going and the smoke rising.

At the other extreme, the Rumfords demand a mason of much skill and thermodynamic savvy. Here it may be best to hold out for a Rumford specialist rather than a good bricklayer willing to attempt a Rumford for the very first time. The design is based on the proven theories of Count Rumford. By the late 1700s, Rumford had becor. e a fashionable doctor of smoky fireplaces throughout Europe. His corrective measures usually consisted of rebuilding the traditional, deep caverns—making them shallow so that the fire sat almost in the room. The back made a tall, gentle forward arc that led to a long, slotlike throat only a few inches wide. Sides flared outward. The result was a fireplace that allowed maximum outreach of radiant energy. To ensure reflectance, Rumford decreed that fireplace walls should not be allowed to collect soot. Details on the construction of Rumfords of various widths are given in Vrest Orton's *The Forgotten Art of Building a Good Fireplace,* from *Yankee,* Dublin, N.H. 03444.

Fireplace specs recommended by the Department of Agriculture are found in *Fireplaces and Chimneys,* USDA Farmer's Bulletin No. 1889, available for 40¢ from the Superintendent of Documents, U.S. Government Printing Office, Washington, D.C. 20402.

Yet for all their glamour and history, open-masonry fireplace styles each offer a compromise. The low, deep design may provide little radiant warmth while

These drawings show rough comparative proportions of masonry fireplaces described in the text.

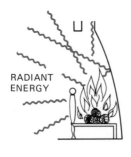
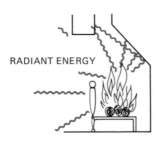
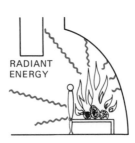

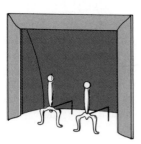
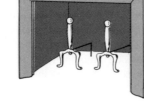
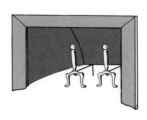

RUMFORD   *CONVENTIONAL*   *CAVERNOUS*

drawing less warm room air. The Rumford, for all its daring engineering and unquestionably superior delivery of radiant energy, may be difficult to build well. And finding a cover that will seal a large-faced Rumford for the night may be difficult. No matter what the design, if the fireplace is open you will lose great volumes of warm room air—far in excess of the fire's needs for combustion.

Construction costs of a masonry fireplace are lowest when the house is built around it. Depending on your locality, the fireplace design, and the reputation of the mason, a new fireplace in a new house may run anywhere from $1200 to several thousand. Costs for adding comparable fireplaces to existing houses will be higher, perhaps even double.

BASIC FIREPLACE CONSTRUCTION GUIDELINES. Again, masonry fireplaces are so inefficient they should not be built as a potential means of home heating. Yet for an occasional evening's mood setter, they're probably no greater polluters than our cars are taking us on unnecessary trips around town. Moderation is the key.

Many do-it-yourselfers with some experience with mortar and stone are intrigued by the prospect of throwing up their own masonry fireplace. They assume that common sense and an ability to lay a brick are sufficient for proper fireplace construction. Many-do-it-yourselfers also see a property-tax advantage in not reporting home improvements to local authorities. *Warning:* Masonry fireplaces, unlike stone walls and outdoor stone barbecues, pose structural and fire hazards unless they are built strictly to code. Here it's imperative to apply for a building permit and satisfy the requirements of inspectors each step of the way.

Support. Floor timbers are not adequate in most instances to support the thousands of pounds that masonry fireplaces weigh. Normally, the fireplace footing should be from the ground up, in no way dependent on house footings or structure for vertical support.

Stone and masonry's low insulating value. Many a house fire has resulted from the transfer of heat through 4 inches of solid concrete. The National Fire Protection Association specifies that fireplaces be at least 8 inches thick. Of this, there must be 2 inches of firebrick. Without firebrick, the walls need to be at least 12 inches thick. Then you need at least 2 inches of fire-stopped space between the outside of the fireplace wall and the house framing.

Hearths for large fireplace openings (6 square feet or more) should extend at least 20 inches out from the opening and 12 inches beyond the sides. Smaller fireplace openings should have a minimum of 16 inches extending in front and 8 inches beyond the sides.

Building the fireplace yourself. Work closely with the building inspector. He'll ensure that your plan results in a solid, fire-safe fireplace, and he or his representatives will make on-site inspections to help you comply with code guidelines.

Then too, you can buy prefabricated fireplaces and wood stoves that offer the option of open flames for fireside enjoyment at less cost and hassle than most masonry fireplaces entail. These units can deliver more heat to the house while drawing less room air up the flue.

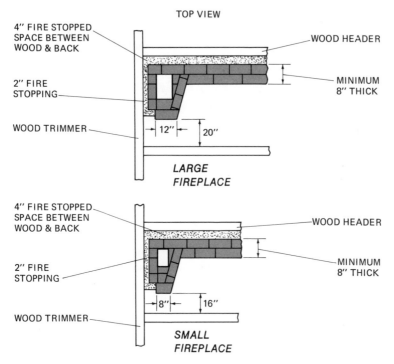

TOP VIEW

4" FIRE STOPPED SPACE BETWEEN WOOD & BACK

WOOD HEADER

2" FIRE STOPPING

MINIMUM 8" THICK

WOOD TRIMMER

12" — 20"

LARGE FIREPLACE

4" FIRE STOPPED SPACE BETWEEN WOOD & BACK

WOOD HEADER

2" FIRE STOPPING

MINIMUM 8" THICK

8" — 16"

WOOD TRIMMER

SMALL FIREPLACE

The National Fire Protection Association recommends these thicknesses and clearances for masonry fireplaces. *Note:* Hearth clearances assume that a spark screen will be used at all times.

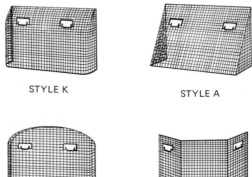

STYLE K

STYLE A

Custom-built spark screens are available from the John P. Smith Company. Here are a few of the most popular styles.

STYLE R

STYLE F

NEW FIREBOX FOR OLD FIREPLACE. The American Stovalator Company contends that 90 percent of the potential heat of a fireplace escapes up the chimney, while the draft siphons off an additional 22 percent of warm room air.

To make masonry fireplaces less wasteful, Stovalator has developed a stovelike firebox that can be inserted into an old hearth. The face of the firebox has a gasketed frame that seals tight against the old fireplace front. Glass doors offer a view of the fire, and combustion air is damped at the base of the firebox. Natural convection draws floor-level room air in at the bottom through a duct that passes behind the firebox and opens into the room again at the top front of the firebox. So air is heated and returned by convection to the room.

Of course, the glass doors greatly limit the release of the fire's radiant energy to the room. But they prevent much of the loss of room air that results from open fires, and some heated convection air is returned to the room. The effect is increased net efficiency over most open masonry fireplaces, and the inlet dampers allow control of the rate at which the wood burns.

Note: There are also ways of plugging wood stoves into old masonry fireplaces, and there are stoves with rear-wall plates that are designed to seal fireplace openings. We'll touch on these in the next chapter.

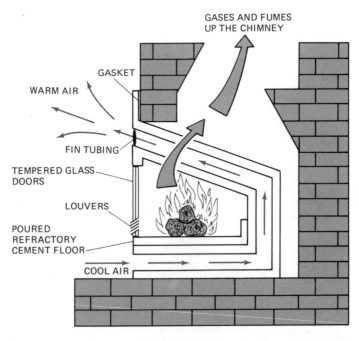

The American Stovalator consists of a glass-doored firebox that is sealed into the old masonry fireplace. Combustion-air intake is limited by louvers. Warm room air is heated and then recirculated by natural convection through fin tubing. No blowers are employed.

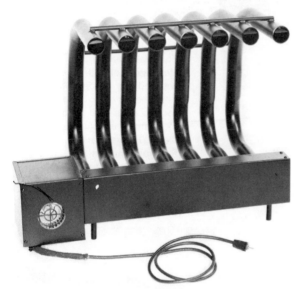

The Aglow Heat-X-Changer (above) and the Shenandoah Fire-Grate are among the many heat reclaimers that are designed to conduct heat from the fire and transfer it to air blown through heat-exchange tubes. For best efficiency, these units should be used in conjunction with a glass screen that reduces the loss of warm room air up the flue.

PREFABRICATED ZERO-CLEARANCE FIREPLACES. These provide the basic look of masonry fireplaces. Yet they often cost under $1000, weigh under 800 pounds, and are advertised for installation by the do-it-yourselfer. The fire chamber resembles a conventional masonry fire chamber and is normally lined with firebrick anyway. The hitch is that the fire chamber is made of heavy-gauge steel enclosed in multiple steel walls spaced for air circulation and for insulation. The insulative value of the walls and bottom results in Underwriters Laboratories approval of placement directly against combustible walls or floor. Since no space need be allowed between the basic unit and combustibles, these fireplaces are popularly called zero-clearance fireplaces.

To install one of these units, you simply follow the manufacturer's instructions. Conventional heavy masonry can be built around the unit, but most people elect to build a wood frame and finish the exterior with real brick, or with brick simulations, or even with wood paneling. The throat of the fire chamber connects to insulated metal flue sections that are normally snap-locked together and then routed to a conventional flue or to a metal-insulated chimney running through the roof or up an outside wall.

This zero-clearance fireplace by Washington Stove Works is typical of the models on the market. Most are lined with firebrick to protect metal and reflect heat. They also contain enough layers of insulation to warrant approval for installation next to combustibles.

This zero-clearance model by Heatilator allows ducting of warmed air in desired directions—even to adjoining rooms. Note that wood framing abuts the fireplace and ducts. Preway, another manufacturer, makes an Energy Mizer, a fireplace that can be rigged to draw only outdoor air for combustion while circulating only warmed room air.

When burned with an open flame, these units should theoretically deliver more heat than masonry fireplaces because they usually offer the same fire-chamber contours as good masonry fireplaces. The bonus is that their ductwork circulates room air, heats it, and carries it by means of natural convection or blowers to the room with the fireplace or to adjoining rooms.

Some of these units come equipped with combustion-air inlets that draw up cool basement air or else air from the outdoors. This colder air may have a quenching effect on the flames and it may even find its way into the house. The intent, though, is to furnish air for combustion and draft and so reduce the flue's demand for warm room air.

Negative notes. This kind of fireplace robs just as much warm room air as masonry fireplaces do unless it is fitted with a tempered-glass screen. We're back again to a fire behind closed doors, for the sake of efficiency.

In earthquake zones, building codes may require special anchoring of fire chamber and chimney sections. And there may be other special requirements. So check with your building inspector before you set up one of these zero-clearance fireplaces, no matter how many laboratories have certified that your unit is a safe one.

Metal fatigue and corrosion should be about your only other worry. Better units usually carry long guarantees in this case, though.

FREESTANDING FIREPLACES. By rights, this heading should apply only to freestanding wood burners with flames open to view. Some people contend, however, that the Franklin-type stove is really a freestanding fireplace. When both doors are open, the Franklin is definitely operating in a fireplace mode. But good Franklins close up tight, and their air inlets and damper provide burns from six to twelve hours. So in the stove mode, the Franklin is not a fireplace. There are many other fireplaces that can be closed and stoves that can be opened. So I guess that our calling these units either stoves or fireplaces depends mainly on how they are used most. Since this is a book on wood heating and not merely on enjoying the visual pleasures of fire, Franklins and other closable units will be covered in the next chapter—on stoves.

Probably the most widely distributed freestanding fireplace is the one with an inverted sheet-metal funnel right over the flames. The funnel's throat connects to single-wall stovepipe before the pipe eventually connects to a masonry or prefabricated-metal chimney. The popularity of these units is based on the appeal of open fires generally, as well as on low cost, ease of installation, and availability in a great variety of designs and decorator colors.

Like open masonry fireplaces, freestanding fireplaces draw tremendous amounts of warm room air up the flue. Their heated metal above the flames and the stovepipe make these units capable of emitting as much heat to the room as most masonry fireplaces. Yet they shouldn't be installed as home heaters unless they can be sealed up tight, with controlled air inlets.

Many cheap and unsafe units have appeared on the market. The building inspector in my locality requires that wood heaters carry the UL label before he'll okay them. That may be more stringent than necessary, but it helps eliminate some of the worst wood heaters.

The more efficient these freestanding fireplaces become, the more they begin to resemble stoves. Heat, efficiency, stoves . . . let's take a look in the next chapter.

Here are three of the most popular styles of freestanding fireplaces. The chain-suspended Duchess (left) and the two-tone Count (below) are by Malm. The closed Franklin with four door panels (bottom) is by the Atlanta Stove Works. Though Franklins can also be operated in an open-door fireplace mode, closed doors greatly reduce the loss of warm room air and help regulate the amount of combustion air reaching the fire.

6 | Wood Stoves

WHETHER YOU BURN wood, oil, or gas, some potential energy will be lost up the flue as heat and unburned volatiles. This results from incomplete combustion as well as from the draft needed to carry off smoke.

Oil and gas furnaces can operate at peak efficiencies of about 80 percent when they are tuned properly and operating continuously. But these same furnaces may operate at efficiencies as low as 40 percent when they are in need of cleaning and a tuneup, and when they must cycle on and off. Well-designed open fireplaces can deliver more than 10 percent of wood's potential energy into the house. But wood stoves of the best quality can do far better. They may deliver 50 to 70 percent of the wood's potential.

These comparative burning efficiencies are important. But it's also wise to consider a fuel burner's net heating effects on a house. Tests conducted by Dr. Jay Shelton show that small closed wood stoves will draw only 22 to 44 cubic feet of warm room air per minute. Larger stoves pull about 33 to 67 cubic feet. Open stoves, such as Franklins, draw from 56 to 110 cubic feet. And, again, fireplaces draw an unforgivable 90 to 330 cubic feet of room air per minute.

Though airtight stoves tend to draw far less room air than other types of wood burners, they do draw a few times more room air than oil or gas furnaces delivering equal amounts of heat. So to compare net efficiencies of wood stoves and fossil-fuel furnaces, be sure to consider losses of room air.

Your house itself will affect the amount of room air lost up the flue. Poorly sealed houses (with daylight showing through cracks) will typically undergo the equivalent of two complete air exchanges with outside air every hour. An air-hungry burner, such as a fireplace, doesn't speed that exchange much because the air is moving freely enough anyway to feed the flue without great effect on the overall exchange. But very tight houses, with only one-half an air exchange each hour, are affected appreciably.

AIRTIGHTNESS AND STOVE MATERIALS. As far as stove construction goes, about the only absolute essential for controlled efficiencies is that the joints be airtight. This allows you to regulate the amount of combustion air by means of movable air inlets. The draft of a non-airtight stove can be limited by a damper in the stovepipe, but leaks in such stoves make low controlled burns difficult to achieve.

There's no single best material for stove construction. You can get high energy efficiencies from stoves made of cast iron, sheet steel, or even ceramic tiles. Nearly all stoves marketed in the U.S. and Canada are made of cast iron or sheet steel or both. The two metals conduct heat equally well. Equal thicknesses of cast iron and sheet steel allow heat to pass from the inside wall to the outside at the same rate. Walls of the heaviest cast-iron and sheet-steel stoves are 3/16 to 1/4 inch thick. Their substantial thermal mass tends to hold heat longer and emit it slower after the fire subsides. The thinner-walled steel stoves heat up quicker and thus are especially popular with people wanting quick heat, such as when arriving at weekend homes. Cast iron is comparatively brittle and can be cracked by the impact of a heavy log. But thinner gauges of sheet steel may warp or bend when the firebox is overheated. So to ensure tight seals for airtightness around doors, makers of thinner-gauge sheet-steel stoves often employ cast-iron doors; cast-iron doors don't need to be protected from warping, they

COMPARATIVE LOSSES OF WARM ROOM AIR

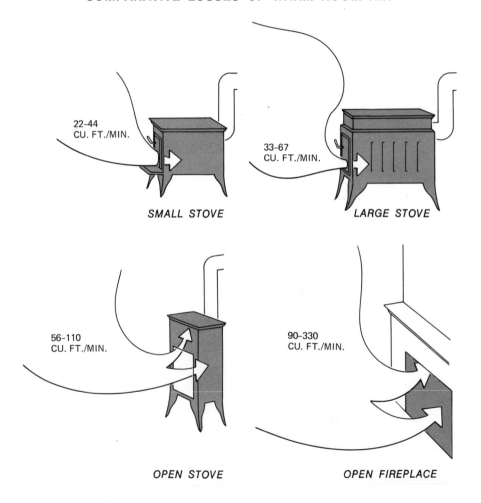

22–44
CU. FT./MIN.

SMALL STOVE

33–67
CU. FT./MIN.

LARGE STOVE

56–110
CU. FT./MIN.

OPEN STOVE

90–330
CU. FT./MIN.

OPEN FIREPLACE

are cheaper to produce in quantity, and they provide a handy means of imprinting the manufacturer's name.

Thickness also has a bearing on the metal's life. Both sheet steel and cast iron are subject to corrosion from acids, and they rust (oxidize) as a result of moisture and intense heat. Extremely thin-walled stoves, such as barrel stoves, may corrode through in only one heating season. But thin-walled stoves can be protected somewhat if you insulate them with layers of sand and firebrick, or with a metal liner.

STOVE TYPES. There are basically two types: radiating stoves and circulating stoves. Radiating stoves are essentially one-layer enclosures that serve both as fire chamber and stove exterior. Most of the heat from a radiating stove is emitted as infrared radiant energy in all directions, warming any surface it strikes. Circulating stoves, on the other hand, have a steel jacket spaced out from the fire chamber. This jacket absorbs most of the radiant energy from the fire-chamber walls and releases heat to air circulating both inside and outside the jacket. The warmed air is buoyant and travels through the house by convection.

If efficiency alone is the factor, you'd have to award equal ratings to the best radiating and circulating stoves. These stoves have attained energy efficiencies of 60 percent and better under home operating conditions. We'll be discussing types of circulating and radiating stoves, as well as specific models, later in this chapter. But for now, let's consider characteristics of the two types.

Circulating stoves. These send about 70 percent of their effective home heat upward by means of natural convection unless the air is redirected by blowers. This is an advantage when the stove is in the basement and you want to direct the heat upstairs immediately. Most circulating stoves are equipped with thermostats that automatically regulate combustion air and heat output. The most efficient circulating stoves need be loaded with wood only two or three times a day; then, in the American tradition, you can "set them and forget them." Also, if you have small children in the house, the jacket of a circulating stove generally won't give the severe burn that the sides of a radiating stove may.

On the negative side, most circulating stoves are designed so that you can't cook on them or keep the tea kettle on. Many of them have an outer jacket that hides the fact that there's a wood stove beneath. Some of these jackets have an artless, tinny look—like many of the oil space heaters have—and the jacket often adds more than $100 to the price. Loading doors are either on top or on the side and so may not let you enjoy an evening fire if you have the urge. In fact, top-loading doors give you a blast of heat in the face and—if you aren't careful—a flame or two.

Some circulating stoves feature a couple of thermostat-controlled air inlets, pull chains, electric blowers, and air and gas ducting. If you love the challenge that any machine with moving parts promises, maybe a circulating stove will prove interesting.

Radiating stoves. This category includes a lot of sorrows and blessings—all essentially single-wall stoves. The best commercially made radiating stoves can match the efficiency of the best circulating stoves: over 60 percent. Yet the worst homecrafted models may be a hazard to operate from the first day, let alone after the first heating season has left pinholes in the walls.

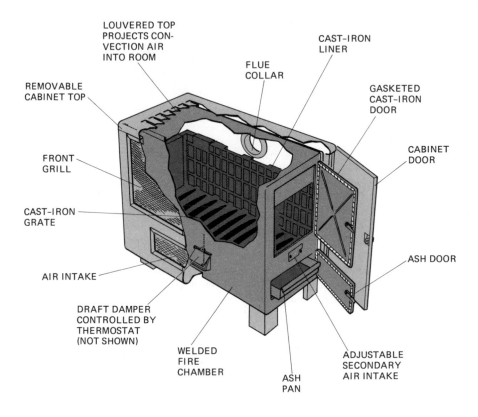

LOUVERED TOP PROJECTS CON-VECTION AIR INTO ROOM

FLUE COLLAR

CAST-IRON LINER

GASKETED CAST-IRON DOOR

REMOVABLE CABINET TOP

CABINET DOOR

FRONT GRILL

CAST-IRON GRATE

ASH DOOR

AIR INTAKE

DRAFT DAMPER CONTROLLED BY THERMOSTAT (NOT SHOWN)

WELDED FIRE CHAMBER

ASH PAN

ADJUSTABLE SECONDARY AIR INTAKE

The Homesteader 240 by Atlanta Stove Works has features common to most circulating-type stoves. Note the double-wall construction designed to convert radiant energy to warm convection air. The thermostat control dial, shown in the photo, determines the damper opening, shown in the drawing. This unit stands 33⅜ inches tall, accepts logs up to 2 feet long, and weighs 240 pounds.

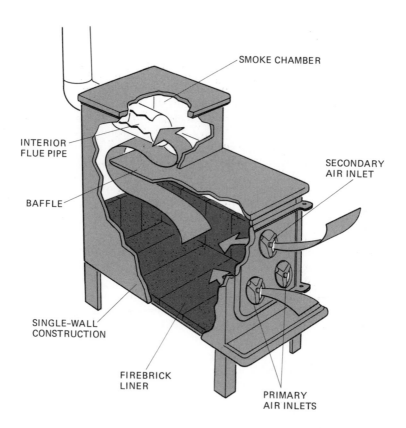

SMOKE CHAMBER

INTERIOR
FLUE PIPE

SECONDARY
AIR INLET

BAFFLE

SINGLE-WALL
CONSTRUCTION

FIREBRICK
LINER

PRIMARY
AIR INLETS

The Huntsman by Atlanta Stove Works is one of a great variety of radiating-type stoves. Some of its features are included in most of the better radiating units. Among these features are secondary and primary air inlets and a smoke chamber that is designed to increase residence time of flue gases and thus promote heat transfer. The Huntsman stands 33½ inches tall, accepts logs up to 2 feet long, and weighs 390 pounds.

Most of the better radiating stoves are made of cast iron. But there are good radiating stoves of heavy-gauge sheet steel too. Joints for sheet-steel stoves are normally welded. Cast-iron contours may be precast, or joints may be tongue-and-groove, sealed with a furnace cement.

Many radiating stoves offer burners for cooking. Some are intended for use either in open "fireplace mode" or in closed "stove mode"—these are especially popular with people who reluctantly had to close up their old masonry fireplaces.

Radiating stoves emit about 70 percent of their effective heat as radiant energy. The energy travels through air without warming it and then heats whatever it strikes. So room air can be comparatively cool even though you may feel quite comfortable because you are absorbing the radiant energy yourself. But this preponderance of radiant energy sometimes makes these stoves unsuitable for use in a basement. Here much of the radiant energy is absorbed by basement walls. But if you like a radiating stove, and it's got to go into a basement, you can build some sort of impermeable screen around it to absorb the energy and let its heat be absorbed by circulating air. This makes the air buoyant, setting up natural convection. In this case, the radiant stove and the screen serve like one large circulating stove or one small furnace.

Among the most popular radiating stoves today are the Scandinavian imports and their American imitators. These are airtights that feature a simple air-inlet system in the door and an airflow pattern that results in good combustion of airborne volatiles and a long residence time of the hot gases. More on this a little later.

Radiating stoves are available in a great variety of styles and even in enameled colors. Cast-iron models often sport cast-in-relief embellishments that increase radiating surface area while adding artful touches. They also strengthen the plates.

OLD MYTHS AND NEW FINDINGS. Recent testing has challenged some of the folk wisdom about stoves. The tests have shown that there are many ways to achieve high energy efficiencies but that lower efficiencies may be more practical at times. Then too, folk concepts often stress the importance of a stove's weight—its thermal mass—but mass may be less important than many other factors. And there's often a big emphasis placed on getting complete combustion, yet it's not reasonable to expect both complete combustion and a high rate of heat transfer through stove walls. Let's look at these points individually.

Energy efficiency. Unlike oil and gas burners that fire intermittently at a set rate and at a constant air/fuel mixture, wood stoves burn with varying amounts of wood of varying qualities and with varying rates of airflow. Heat output is regulated by varying the amount of wood, or air, or both.

To increase room temperature from a given load of wood, you have to provide more combustion air. This fans the flames, creating higher temperatures inside the firebox and increasing the draft. The hotter firebox sends more heat to its outer walls and increases room temperature. But the increased draft also draws off higher-temperature flue gases—wasting more of the wood's potential energy than if the wood had been burned at a slower rate.

So, if you burn wood fast, you'll achieve higher room temperatures. But you'll also lower the energy efficiency because more of the wood's potential heat is drawn up the flue.

Stove weight. The weight—the mass—does not determine the amount of heat delivered to the room. Rather, it determines the speed with which the mass of a stove will gain heat and then transfer it to the room. That's why thin-walled stoves will begin heating a room sooner after startup than thick-walled stoves will. But thick-walled stoves offer a steadier rate of heat when the amount of heat inside the firebox fluctuates. This steadiness is the result of the storage capacity of the greater mass of metal.

It's possible to build two stoves with identical interiors and airflow patterns that would deliver about the same amount of heat even though one stove might weigh only half as much as the other. The heavier stove would probably offer a steadier rate of heat, but the overall amounts of heat in Btu could be almost identical.

Weight alone plays a major role in steady heat output, but indirectly it usually tells you that a stove offers greater amounts of surface area for heat transfer to the house. In a given model line by one manufacturer, the heavier stoves are normally capable of greater heat outputs than the lighter ones, mainly because of their greater surface area and larger fire chambers. But these larger stoves also require more wood in order to show their stuff.

The compromise: complete combustion vs. heat transfer. If you can burn all volatile gases and mists before they exit up the flue, you'll have complete combustion. Emissions of an ideal, complete burn are simply carbon dioxide and water vapor. The trouble is, for complete combustion you need an ideal mix of air and volatiles at all parts of a fire in all stages of its life, and the mix must occur at a temperature of 1100°F or better for ignition.

In the best-engineered radiating and circulating stoves, air inlets, airflow patterns, and fire-chamber size are designed to achieve fairly complete combustion when the stove is set to turn out moderate to high amounts of heat. Here there's need for excess volumes of combustion air to ensure that the air reaches most places where volatiles and high temperatures exist. Yet this increases draft and carries some unburned volatiles and much heat up the flue. So to get complete burns, you normally have to lose more heat up the flue. The heat in the exhaust gases simply travels so fast that there's little chance for it to be absorbed by the stove or stovepipe for transfer to the room.

You can reclaim some of this heat by using a longer stovepipe. This increases the residence time of hot gases and allows time for more complete transfer of heat—in this case through stovepipe walls. But when the stove is run at low power, this also lowers flue temperatures enough to favor creosote buildup. Some stoves have a built-in smoke chamber, apart from the fire chamber itself, to improve heat transfer. Double-drum stoves, with one drum serving as fire chamber and the other as smoke chamber, achieve the same purpose; that is, hot gases are exposed to more heat-transfer surfaces for longer periods.

Even though it's smart to strive for relatively complete combustion, you shouldn't strive for 100 percent heat transfer, because that can result in insufficient draft. For most stoves, you need a heat loss of at least 20 percent up the flue in order to draw adequate amounts of combustion air for the fire and to prevent smoke from entering the house.

There's more. For long, steady burns of eight to twelve hours, you have to limit the amount of combustion air entering the fire chamber. This promotes a slow, steady burn and a slow, low-volume draft up the flue. For these low, long-lasting fires there's just not enough combustion air, nor are there high enough temperatures everywhere, to ignite all the gases and tar mists. Yet these hot volatiles circulate and exit through the stove and stovepipe slowly enough to allow good heat transfer to the room.

At these low rates of burn, combustion tends to be less complete than when air inlets are wide open. Yet there's a far higher percentage of heat transfer per pound of wood consumed. This, then, is the fundamental compromise between combustion and heat transfer.

REPUTED CREOSOTE PRODUCERS. Incomplete combustion sends volatile gases and tars up the flue at relatively low temperatures of 150°F to 500°F. These temperatures encourage condensation of water and volatiles that make up creosote. That's why many of the top-rated stoves of both the circulating and radiating types also carry stigmas of being creosote producers. Don't blame the stoves. They're just responding to the way they are operated.

If you want to avoid creosote buildup to the greatest extent possible, you should avoid slow, steady burns. Yet faster burns may give you heat faster than you want it, say, overnight. And faster burns consume more fuel. Anyway, creosote isn't such terrible stuff if you clean your stovepipe and flue often enough to prevent much accumulation.

AIRFLOW INSIDE STOVES. Airflow through a stove plays a vital role in determining completeness or incompleteness of combustion, as well as the amount of heat transfer. Factors affecting this are (1) size and number of air inlets and (2) internal channels that direct flow. Here are some basic types of flow.

Updraft. Old-style stoves, particularly the potbellies, employ a simple updraft that enters at the base of the stove and passes through the wood to a stovepipe at the stove top. There is usually an inlet door above the wood to admit secondary air and help ignite unburned volatiles. The result can often be a fairly complete burn, but with flames lapping into the stovepipe. Here the stovepipe may be so hot that it can support combustion itself. This is why red-hot stovepipes and potbellies are commonly associated.

Antique and novelty values of potbellies and tall updraft stoves are generally higher than heating value, unless the stove design somehow provides for longer residence time of the hot gases. This residence time determines heat transfer to the room before flue gases escape. Of course, the old method of increasing heat transfer was to employ a long stovepipe. But that's just not safe.

Diagonal flow. This is the flow common to simple box and barrel stoves. It begins at a vent low in the stove front and sweeps diagonally upward through coals, wood, and flame and then out the stovepipe at the upper back of the stove.

Cheaper cast-iron and sheet-steel box stoves often have air leaks that create haphazard turbulence that may mix air and unburned volatiles well enough to yield fairly complete secondary combustion at some draft settings. But the leaks also admit great amounts of excess air and thus rob more room air than an airtight stove of similar size would. The leaks also make low, controlled burns almost impossible to achieve without the aid of a stovepipe damper and some luck.

The basic deficiency of diagonal flow in an airtight stove is that it doesn't provide enough mixing of secondary air with hot volatiles for complete combustion. Besides that, the hot gases have only a brief residence time and streamline near the center of the flow without coming in contact with metal before they are drawn up the flue. Short residence time restricts the amount of heat transfer through stove walls into the room.

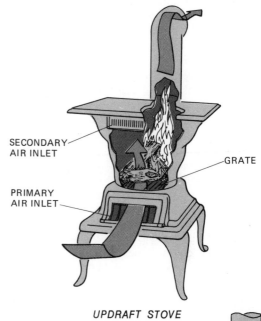

SECONDARY
AIR INLET

GRATE

PRIMARY
AIR INLET

These are air-flow patterns of stoves that don't allow much residence time of flue gases. Residence time and gas-metal contact are desirable because they promote heat transfer to the stove walls and the room.

UPDRAFT STOVE

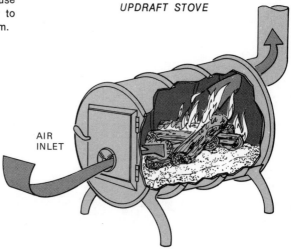

AIR
INLET

DIAGONAL FLOW

S-flow and variations. Probably the most sensational airflow trend in North America in recent decades involves the S-flow long employed in Scandinavian stoves. These Scandinavian radiating stoves have enjoyed such general acclaim and strong sales figures that American imitators have cropped up.

In an S-flow stove, a simple, manual air inlet in front delivers air to the near end of the logs lying along the stove's long axis. An iron baffle plate provides a horizontal partition just inches above the logs and absorbs much heat while reflecting and radiating much more back onto the logs. This, together with baffle plates on the sides, bounces a lot of heat inward. Here you have a tight-fitting, hot combustion chamber.

As inrushing air meets the fire, the turbulent air and volatiles mix, and the hot fire chamber causes ignition. But since the overhead baffle plate prevents hot gases from escaping directly out the back of the stove, the gases follow the path of least resistance under the baffle plate to its front edge, loop over the top of the baffle plate, and then head back toward the stovepipe—and away.

All this congestion and turbulence and ignition takes time—time enough to allow good amounts of heat transfer to baffle plates and stove shell, time for relatively complete combustion when air inlets are set for moderate and high rates of burn.

Variations on this S-flow may include a separate smoke chamber rather than a mere baffle-plate partition. Here the stove may have a fire chamber on the bottom and a smoke chamber just as large above it. Double-drum stoves employ this principle and have been popular projects for do-it-yourselfers.

Caution: At extremely low, steady rates of burn, these stoves may transfer so much heat that flue gases rise at low, creosote-condensing temperatures. I've had no difficulty getting nine to ten hours of heat-producing burns on single overnight loads of wood with my Jøtul 602. Dr. Jay Shelton, a leading authority on wood heat, conducted a test in which a Jøtul 602 yielded an average of 4000 Btu per hour for the first sixteen hours on one load of wood and then declined in output gradually from 2000 Btu to zero at the end of *twenty-six hours.* But most people don't want to bother with burning at 4000-Btu heating rates. That

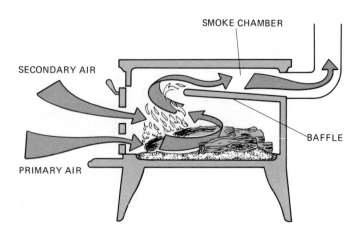

SMOKE CHAMBER

SECONDARY AIR

BAFFLE

PRIMARY AIR

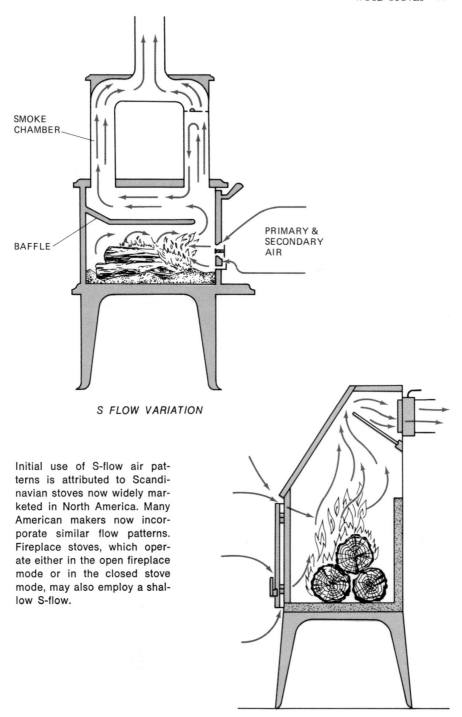

SMOKE CHAMBER

BAFFLE

PRIMARY & SECONDARY AIR

S FLOW VARIATION

Initial use of S-flow air patterns is attributed to Scandinavian stoves now widely marketed in North America. Many American makers now incorporate similar flow patterns. Fireplace stoves, which operate either in the open fireplace mode or in the closed stove mode, may also employ a shallow S-flow.

S-FLOW IN FIREPLACE STOVE

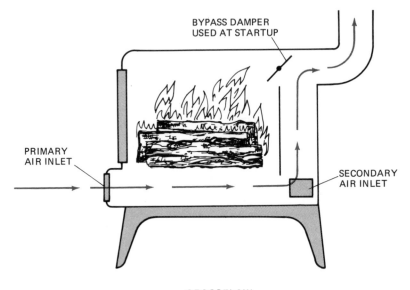

CROSSFLOW

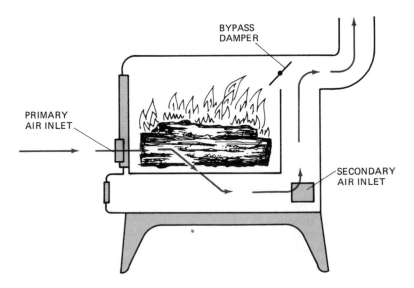

DOWNFLOW

For both the crossflow and the downflow stoves, complete combustion depends on the effectiveness of the secondary air inlets in igniting unburned gases. At low rates of burn, these gases may often have cooled too much to ignite when mixed with secondary air. In this case, electric blowers and moderate to high rates of burn may be necessary in order to obtain secondary combustion.

same little 115-pound stove gives its best energy efficiency per pound of wood at about 15,000 Btu per hour. It delivers its lowest efficiency at its maximum output of about 27,000 Btu per hour.

Larger stoves employing S-flow and variations with smoke chambers offer their best energy efficiencies at about 30,000 Btu per hour and lowest efficiencies at top outputs of about 50,000 Btu per hour.

Downflow and Crossflow. In these designs, primary air enters either above the logs or at one side and is passed through the logs. Then a secondary air inlet below the coals (downflow) or behind the coals (crossflow) is supposed to ignite hot volatiles that would otherwise escape up the flue.

In one downflow model, the DownDrafter, the system doesn't work at peak efficiency without an electric blower. I've never used a DownDrafter, but I've read that they can be balky unless the flue has a good natural draft. Here the blower and the draft are controlled by a thermostat that senses flue temperature.

Whether downflow or crossflow, stoves that depend on secondary air inlets for complete combustion are under challenge. Dr. Jay Shelton contends that secondary air inlets located somewhat distant from the fire chamber often may fail to ignite unburned volatiles. This failure occurs mainly during moderate to low rates of burn. In this case, the unburned volatiles cool below their 1100°F ignition temperatures before the secondary air reaches them. And without ignition, the secondary air acts like a leak, increasing the loss of warm room air.

PREHEATING OF AIR. Some stove designs route combustion air through pipes or ducts so that it will absorb some stove heat before it mixes with volatiles. This preheating may amount to only a couple hundred degrees under favorable

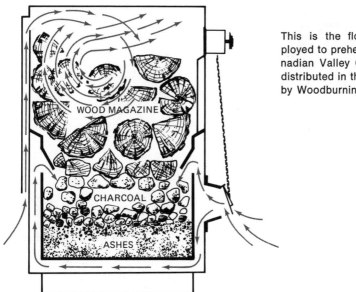

This is the flow pattern employed to preheat air in the Canadian Valley Comfort Heater, distributed in the United States by Woodburning Specialties.

conditions. And, of course, every little bit helps. But remember that most volatiles ignite at temperatures of about 1100°F. So that preheated air still is far from contributing much. And to gain heat, the entering air needs to draw heat from the stove that would otherwise be entering the room.

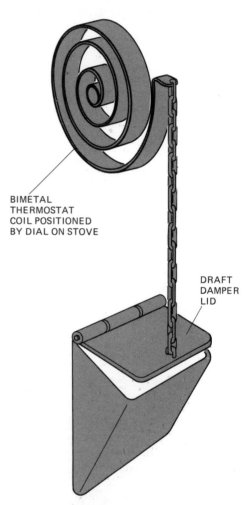

BIMETAL
THERMOSTAT
COIL POSITIONED
BY DIAL ON STOVE

DRAFT
DAMPER
LID

With a dial-type air inlet, a simple quarter turn reduces the flow from wide open to fully closed. Secondary air enters essentially through the topmost of the openings and primary air through the bottom.

In bimetal thermostat assemblies, two different metals are bonded and then coiled. These metals respond to temperature changes by expanding and contracting at different rates. This causes the end of the coil to move, opening and closing the damper lid.

THE KEY FACTOR. *The most crucial factor for complete, efficient burns is high temperatures where the unburned volatiles are traveling and mixing with air. Design and airflow that promote and exploit hot fire chambers tend to get the most reliable secondary burns—and the most complete combustion.*

AUTOMATIC VS. MANUAL DRAFT CONTROL. The choice is often made for you when you select either a radiant-type or circulating-type stove. Most radiant stoves offer only manually operated air inlets. Circulating stoves usually have one or two thermostats that are designed to let you set the thermostat when you load the fire chamber and then forget about controls until it's time to add more wood.

Thermostatic control is easier than manual. In some ways it's a lot like keeping the family dog in a kennel, feeding him twice a day, removing his waste, patting him on the head, and then neglecting him until the next feeding. On the other hand, manual control is more like having a house dog, responding to his requests for walks, talking to him more often, and generally developing a closer relationship with him. This, of course, takes more time, but many people find it rewarding. And so it can be with stoves. However, if you own a dog and decide to get a stove too, you may have to make some hard choices.

A manual-draft stove doesn't really need much attention anyway. Normally, an infrequent tap of the finger is all that's needed to slightly open or close the inlet for higher or lower rates of burn. Once you get to know your stove and become attuned to its whispers, you can perform the "tap" almost without missing a step as you move about the house attending to other matters. Often, though, the stop may become a habitual, self-indulgent one. You may find yourself kneeling down and opening the door, knowing the fire's doing well but checking it anyway—like patting the dog when you pass him.

For automatic control, bimetal thermostats are used. For this, two strips of different metals are bonded together in a coil (like a watch spring) and connected by a chain to an air inlet lid. Temperature increases and decreases near the coil cause the metal strips to expand and contract at different rates. This tightens and loosens the coil, tensioning and releasing the chain—and opening and closing the air lid.

Some lids have magnets that cause the lid to snap shut and pop open in response to the chain. This has an interesting effect on heat output. It tends to make the stove produce sharp, but brief and almost unnoticeable, heat increases and decreases. Lids without magnets tend to close and open more gradually, and they consequently produce more gradual heat increases and decreases. Here heat peaks and valleys may be slightly more noticeable than with the more frequent magnetic cycling.

Mechanically, the thermostat is probably the most vulnerable part on most stoves. If it somehow becomes clogged, will it function properly? If a chimney fire starts, will inrushing air cool the thermostat, causing it to hold the lid open rather than closing down and limiting combustion air to the chimney fire? In short, the location of the thermostat and its protection from clogging deserve a thought.

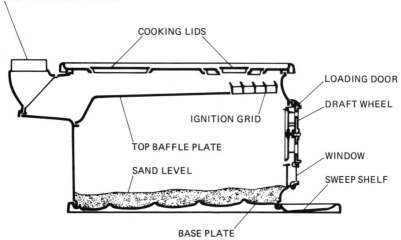

ROTATING FLUE FITTING

COOKING LIDS

LOADING DOOR

DRAFT WHEEL

IGNITION GRID

TOP BAFFLE PLATE

SAND LEVEL

WINDOW

SWEEP SHELF

BASE PLATE

IGNITION GRID

ADJUSTABLE SIDE BAFFLE PLATES

The American-made Cawley/ LeMay 600 and its smaller sibling, the 400, incorporate the best features of the Scandinavian stoves, while adding a few refinements. An ample sweep shelf catches renegade ash and embers. A 35½-inch-high stove top requires less stooping and crouching. The top has a raised edge to catch spills and prevent cookware from sliding off. The 600 stove accepts 2-foot logs and weighs 385 pounds. The 400 takes 16-inch logs and weighs 300 pounds.

The StoveWorks' Independence is a creation of Larry Gay, wood-heat author and inventor. This steel stove offers a baffled S-flow of air. The soapstones are optional. Spaced ½ inch from the stove walls, they convert radiant energy to warm convection air. The stove stands 31 inches tall, accepts 28-inch logs, and weighs 285 pounds with soapstones. An Independence Jr. is available with a firebox that is half the size. Both models offer hot-water coil options.

The Model 30 barrel stove by Yankee Woodstoves is available complete, or as a kit. A horizontal baffle promotes the S-flow draft pattern. If you have your own barrel, you can buy parts separately, including legs, cast-iron door, flue outlet, baffle plate, and (not shown) a top-mounted cooking plate and an ash apron.

The Lange Model 6203 BR from Scandinavian Stoves, Inc., has a cast relief patterned after European tile stoves. A horizontal baffle results in an S-flow draft. The stove stands 41 inches high at the top of the flue fitting. It accepts 16-inch logs and weighs 213 pounds. Lange also makes many other models.

→

Better 'n Ben's fireplace stove, available through C&D Distributors, is installed in the old fireplace opening. The stove box measures 18×18×24 inches and is made of 11-gauge steel. Back panels are available in all sizes. Approximate shipping weight, complete, is 164 pounds. Many people install a stove in an old fireplace opening by running the stovepipe into the fireplace throat and sealing around the pipe with mineral wool.

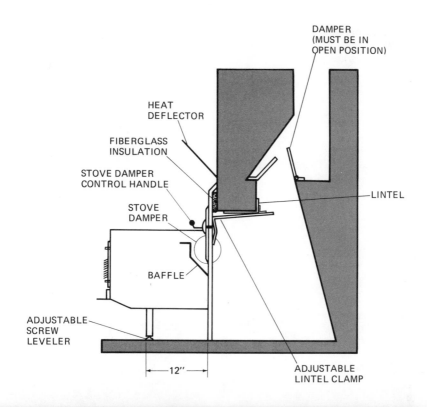

DAMPER
(MUST BE IN
OPEN POSITION)

HEAT
DEFLECTOR

FIBERGLASS
INSULATION

STOVE DAMPER
CONTROL HANDLE

STOVE
DAMPER

LINTEL

BAFFLE

ADJUSTABLE
SCREW
LEVELER

12″

ADJUSTABLE
LINTEL CLAMP

The Jøtul Combi-fire No. 4, distributed by Kristia Associates, is a sensible choice if you like occasional open fires but want the efficiency of an airtight stove most of the time. The bottom panel swings down to offer a fireplace mode, and it closes tight as a bank vault; with air entering only at the adjustable inlet in the door. The No. 4 stands 41 inches tall, accepts 16-inch logs, and weighs 299 pounds. There are many other Jøtul models.

The Kota C sauna stove is imported from Finland by Solar Saunas. This basically cast-iron model weighs 265 pounds without the rocks. A steel model, the Kota L, weighs 165 pounds. Both can be fitted with an 8-gallon water-heating tank. Also available are plans for sauna construction and solar hookups.

7 | Multi-fuel and Wood Furnaces

WHEN IS A STOVE not a stove? When it's a furnace.

Stoves are basically one-room or multi-room heaters. They warm the house by emitting radiant energy and by producing warm convection air. Rooms, open doorways, and stairwells serve as ducts for stove heat. Stoves simply heat surrounding air and hope for the best.

Furnaces, on the other hand, employ either heat exchangers connected to air ducts or hot-water boilers connected to pipes. Using air or water, furnaces distribute the heat according to your wishes.

There are also capacity differences between stoves and furnaces. Few stoves offer outputs of over 50,000 Btu per hour, which is what you need to heat three to six rooms in northern winters during the coldest months. Small furnaces normally offer at least 50,000 Btuh, and the large wood boilers produce up to 350,000 Btuh, enough for a very large house indeed.

Stove and furnace heating capacities have a direct relation to their demands upon a household labor force. No matter where a stove is located, tending it amounts to a parlor pleasantry, after which you can pull up a chair and toast your toes. But furnace tending is really a boiler-room chore. It involves loading 4-to-8-inch logs that may range from 2 feet long for some furnaces up to 5 feet long for the Longwood furnace.

Let's say you want to pull 100,000 Btu per hour from a furnace. If the furnace operates at 50 percent efficiency, you have to generate a 200,000 Btuh output from the wood. It's safe to figure about 7000 Btu from a pound of seasoned hardwood. So you'd have to burn about 28½ pounds of wood an hour (200,000 ÷ 7000 = 28.6) to deliver the 100,000 Btuh. Over a 24-hour period, at that rate, you'd be putting 686.4 pounds of wood through the furnace.

Fortunately, many houses don't ever need 100,000 Btuh during the winter, and at night, lower thermostat settings can reduce the wood requirements greatly. Yet, this example gives you an idea what kind of boiler-room workload you'll be in for if you fire wood, full throttle, around the clock in one of the high-capacity furnaces. And, remember, burning a lot of wood involves more than boiler-room chores. You'll do more cutting, splitting, and hauling—and more frequent flue cleaning. If you own a large woodlot and do the work yourself, the fuel is practically cost-free. Under ideal conditions, you'd need about ten acres for a yield of eight to ten cords of wood a year forever. This would probably be fair average consumption for a medium-size house, in a northern state, fueled almost completely by wood. If you buy eight to ten cords at $50 a cord, the wood would cost you $400 to $500 annually.

WOOD-ONLY FURNACES AND BOILERS. These may be used as the sole source of central heat, but more often they serve as helpers connected to a conventional oil or gas furnace and its distribution system.

Hot-air models have fire chambers enclosed in a sheet-metal jacket with a hot-air plenum that may connect to the hot-air plenum of the old oil or gas furnace, or the hot-air plenum of the wood furnace may connect to the other furnace's cold-air return and thus preheat air to the other furnace. And the other furnace's cold-air return may duct into either furnace in response to thermostatically controlled duct doors.

These units are strictly hand-fired, such as those by Kickapoo, Wesco, Charmaster, Ram Forge, StoveWorks, Riteway. (See Appendix for addresses.) To operate them in tandem with the other furnace, you simply set the thermostat for the other furnace a few degrees lower than the setting on the thermostat for the wood furnace. So when wood runs low and the heat output drops, the other furnace fires, automatically supplying just enough heat to satisfy its own room thermostat. All the while the wood is burning down to ash, it delivers heat in decreasing amounts, and the other furnace increases its firing as necessary. Once the wood fire is completely out, the other furnace operates alone.

Some hot-water boilers burn only wood. Others may burn wood and coal. Like wood hot-air furnaces, wood boilers can be thermostatically regulated for tandem use with a gas or oil furnace.

Methods of hookup vary, depending on the model. The accompanying drawings show a couple of methods. Makers include Wesco, Carlson, Combo, Riteway, StoveWorks, HS Tarm, Ram Forge, Hoval, and Tasso. Carlson, Combo, Ram Forge, StoveWorks, and Riteway handle their own sales. Tassos are distributed by Integrated Thermal Systems. Wescos are available from Wood Energy Systems. Hovals are available from Arotek. (See Appendix for addresses.)

Advantages. There are several advantages in operating a wood furnace in tandem with the old furnace:

• Costs for the wood furnace and installation may be under $1000 if you can continue using your old furnace while sharing its chimney.

• Costs for additional ductwork or water pipes are reasonable because you can plug right into the existing network.

• Building inspectors may be favorably inclined toward this means of heating with two fuels, because there are no worries about gas or oil burners sharing the combustion chamber with wood or coal, as is the case with multi-fuel furnaces.

Disadvantages. The two major disadvantages of using tandem furnaces are these:

• They require more floor space than a single multi-fuel furnace. Essentially, you need enough room for two furnaces and hardware that allows shortest possible connections to the chimney.

• Local codes or a too-small flue in your present chimney may require a second chimney for the wood furnace. So you may have to weigh the practicality of adding another chimney against that of buying a multi-fuel furnace for use with your present chimney.

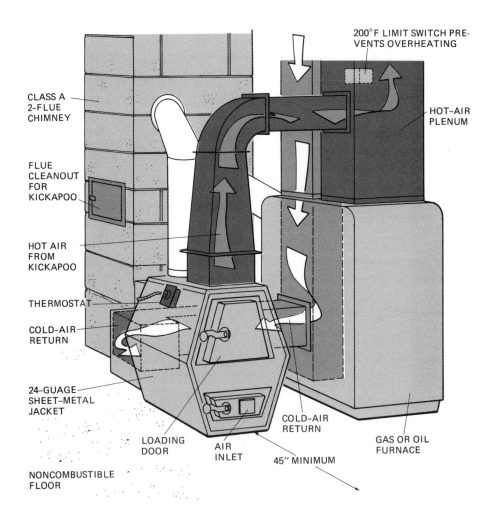

200°F LIMIT SWITCH PRE-
VENTS OVERHEATING

CLASS A
2-FLUE
CHIMNEY

HOT-AIR
PLENUM

FLUE
CLEANOUT
FOR
KICKAPOO

HOT AIR
FROM
KICKAPOO

THERMOSTAT

COLD-AIR
RETURN

24-GUAGE
SHEET-METAL
JACKET

LOADING
DOOR

AIR
INLET

COLD-AIR
RETURN

45″ MINIMUM

GAS OR OIL
FURNACE

NONCOMBUSTIBLE
FLOOR

At left is a wood-only Kicka-poo Home Furnace installed in tandem with a gas or oil fur-nace. Note that the furnaces enter separate flues in a two-flue chimney. Local codes may allow using only one large flue for both units, even though the National Fire Protection Asso-ciation advises against it. Kick-apoo Stove Works makes this unit, shown at right, which stands 35¼ inches tall and weighs 334 pounds, as well as the Parlor Furnace weighing 364 pounds and two stoves each weighing about 250 pounds—all having the basic slope-sided design.

MULTI-FUEL FURNACES. Although multi-fuel furnaces have been popular in Europe for decades, they didn't get much notice in North America until the mid-1970s. It simply took a few years after the 1973 oil embargo for some North American makers to increase production and for others to tool up for pro-duction and enter the market.

As a result, some inventive heating contractors have become company presi-dents. The owner of a sheet-metal shop suddenly needed two shifts for his newly formed furnace factory. Long-established companies found their sales figures soaring.

Nearly all multi-fuel furnaces for hot-air and hot-water systems are fueled either by gas or oil in league with wood, or in league with wood and coal. Lynndale offers one wood hot-air furnace that can be used with an electric heat pump. In Combo, Riteway, Charmaster, Bellway, and Longwood furnaces, the burner for the fossil fuel fires into the same combustion chamber that holds the wood. Other makes, such as Hoval, Duo-matic, Longwood, Lynndale, Dual Fuel's Newmac, and Wilson Industries' Yukon, have separate combustion cham-bers for the wood and the oil or gas. (Charmaster's Model II has glass doors on the side that allow a view of the fire, and they open wide for optional fireplace-mode operation. So the furnace itself can be installed behind a partition wall, with only the screened fireplace in view.)

Multi-fuel furnaces can run solely on gas or oil, and they can augment the wood whenever it isn't turning out enough heat to satisfy the thermostat. Some makers warn that burners should never be fired directly into tightly packed coal for fear of explosions. Leo Graves, owner of Comba Furnaces, says he'll use only oil in combination with wood, because he feels a gas/wood furnace could mal-

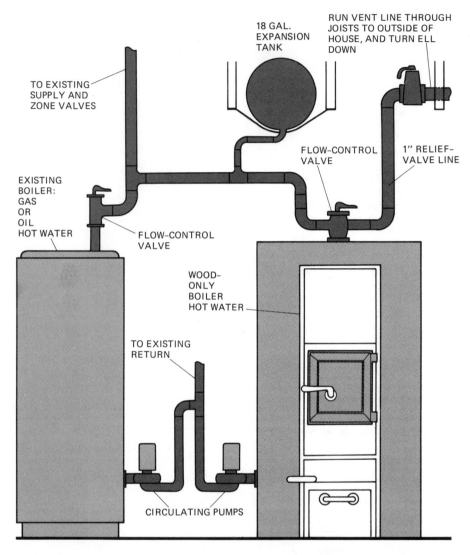

18 GAL.
EXPANSION
TANK

RUN VENT LINE THROUGH
JOISTS TO OUTSIDE OF
HOUSE, AND TURN ELL
DOWN

TO EXISTING
SUPPLY AND
ZONE VALVES

FLOW-CONTROL
VALVE

1" RELIEF-
VALVE LINE

EXISTING
BOILER:
GAS
OR
OIL
HOT WATER

FLOW-CONTROL
VALVE

WOOD-
ONLY
BOILER
HOT WATER

TO EXISTING
RETURN

CIRCULATING PUMPS

This hot-water network is used for Combo's wood-only boiler when it operates in tandem with an existing gas or oil boiler. Circulating pumps are thermostatically controlled. Aquastatic limit switches admit water into the supply at a recommended 180°F. To start up the unit you switch on a gun-type oil burner that ignites the wood. Then when the wood is hot enough, the burner shuts off. Two models are available with outputs of 84,000 and 126,000 Btu.

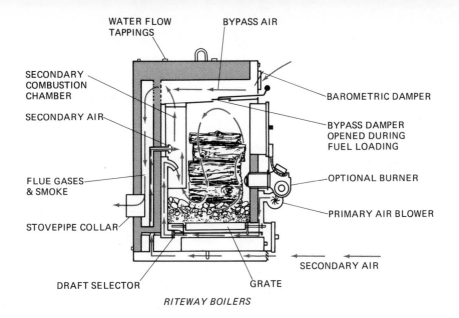

WATER FLOW TAPPINGS

BYPASS AIR

SECONDARY COMBUSTION CHAMBER

SECONDARY AIR

FLUE GASES & SMOKE

STOVEPIPE COLLAR

DRAFT SELECTOR

BAROMETRIC DAMPER

BYPASS DAMPER OPENED DURING FUEL LOADING

OPTIONAL BURNER

PRIMARY AIR BLOWER

SECONDARY AIR

GRATE

RITEWAY BOILERS

RETURN AIR

HOT AIR

BYPASS-AIR FLUE

EXCHANGER PREHEATS RETURN AIR

BAROMETRIC DAMPER

BYPASS DAMPER OPENED DURING FUEL LOADING

STOVEPIPE COLLAR

OPTIONAL BURNER

PRIMARY AIR BLOWER

RETURN-AIR BLOWER

SECONDARY AIR

PRIMARY-AIR PORTS

SECONDARY COMBUSTION CHAMBER

GRATE

SECONDARY AIR

RITEWAY FURNACES

Riteway furnaces and boilers burn wood or coal exclusively or, optionally, in the same fire chamber with gas or oil. For multi-fuel burning, the thermostat for the oil or gas burner is set a few degrees lower than the thermostat for the wood or coal. Riteways are equipped with large ash pans that require cleanout only about once a month. Riteway offers three hot-air furnaces with maximum outputs ranging from 160,000 to 350,000 Btu, and there are four boilers with top outputs ranging from 125,000 up to 350,000 Btu.

The Longwood Dualfuel burns logs up to 5 feet long. Most other multi-fuel furnaces have burners that start up the wood fire and then shut off, letting the
wood consume itself before taking over. But in the Longwood, the oil or gas
burner fires the wood periodically within a no-draft fire chamber. This lack of
draft makes the wood burn as relatively flameless charcoal. Each call from the
thermostat starts the burner. When the thermostat senses enough heat, the
burner shuts off and the charcoal continues to burn, gradually dying down until
the thermostat starts another firing. Longwood claims an output of 150,000 Btu
with a 100-pound charge of wood burning with gas or oil over a 12-hour period.
But remember, this unit is guaranteed to use significant amounts of gas or oil,
unlike other makes that fire only when they run low on wood.

function, leaving a gas pocket in the bottom of the unit, which could result in
an explosion. Some makers do offer gas/wood options, however.

Many of the multi-fuel furnaces may cost only $1000 more than comparable
single-fuel furnaces. So if the old furnace is ready for the scrap yard anyway,
a multi-fuel furnace would be no more expensive than installing a new oil or gas
furnace with a wood-furnace helper, and if a second furnace would require an
extra chimney, you'd be saving installation costs by going with a multi-fuel unit.

Code taboos. The largest single obstacle for makers of multi-fuel furnaces has
been local codes. Some furnaces are listed by respected European and Canadian
testing labs, and many furnaces carry parts with U.S. listing by the Underwriters

Laboratories. But if these units don't carry complete UL listing, some local officials may not approve them for installation. So check before you buy.

The simplest way to find a multi-fuel furnace acceptable to local officials is to search for ads in newspapers covering your home area. Or you can check the Yellow Pages under headings such as "Heating," "Oil," or "Gas." Many fuel companies with an eye toward profits sell their own furnaces. So your own fuel company may offer a line of multi-fuel furnaces. Anyway, you can bet that local dealers in multi-fuel furnaces will have gone to bat for approval from local officials. Get the dealer's assurances, and then check with local officials just to be safe.

If you can't obtain any encouragement locally, you should write some of the companies mentioned in this chapter. Addresses are listed in the Appendix.

Maintenance caution. No two makes of multi-fuel furnaces are exactly alike. Some makers claim that their units burn wood so clean that you'll hardly have a worry about creosote. Most others frankly admit that heat exchangers or boilers should be checked weekly or biweekly and that the interior should be cleaned *at least once a month.* In fact, Combo employs such a highly effective heat exchanger that it condenses most of the creosote that would normally accumulate in a chimney flue. Combo's warranty is void if damage is due to your failure to clean the heat exchanger. That warning may scare off some customers, because creosote can corrode metal and because creosote and ash buildups can insulate the walls of boilers and heat exchangers enough to reduce heat transfer. Yet Combo's warning has intriguing implications. The Combo is transferring so much heat that volatiles are cooling enough to condense before they reach the chimney. That means less heat is escaping up the flue. Provided the Combo gives fairly complete combustion, the result should be high home heat value for the amount of wood consumed.

Furnace makers that boast of low creosote accumulations in heat exchangers and chimney flues (due to high temperatures) may really be reducing creosote by sacrificing potential home heat up the flue. Ask about this trade-off before you make a purchase.

Multi-Fuel consumer reminders. Before you buy one of these units, ask yourself the following questions:

• Can I get adequate service out of my present furnace in league with a wood stove or wood-only furnace? If so, costs may be less.

• Will local officials approve the multi-fuel unit I'm interested in? And if so, what will be the hidden installation costs? Rebuilt chimney? Second chimney? Ducting? Plumbing and control devices?

• Will I have a sufficient supply of wood to make the new furnace worthwhile? If so, who in family will be able to stoke logs 2 feet long? 4 feet long? 5 feet long?

• Who will check the unit for creosote weekly? Who will clean the unit at least monthly?

8 | Installing the Wood-burning System

It's EASY TO OBTAIN a good wood heater these days. Just consider the advice in previous chapters and that of your wood-burning friends, before consulting the Yellow Pages. Then it's a matter of picking the model that seems to meet your needs.

The real challenge lies in the installation. Sure, the installation affects your home's appearance and comfort levels. But most important, the installation determines fire safety. Without that, the rest just isn't worthwhile.

First, you'll want to pick a good location for your wood heater. If you are buying a wood-burning furnace, you'll probably put it next to your oil or gas furnace, or simply replace the old furnace with a new multi-fuel furnace. If you are buying a wood stove, your options may be many. You'll want to weigh the basic factors and decide which apply and which deserve priority. Those that follow should give you a start.

- Can you use an existing chimney? If so, it will mean placing the stove nearby.
- Which rooms are most important to heat? Here, remember that wood stoves distribute heat best when they are located low and centrally, in relation to the prime rooms.
- Can you take advantage of an open stairwell or air grates for heat distribution?
- Is there an outside door near enough to make wood hauling convenient?
- Will the stove disrupt normal traffic patterns? Never skimp on the firesafe clearances explained at the end of this chapter. You'll be better off finding a stove location that won't disrupt traffic, rather than skimping on clearances to accommodate traffic.
- Will the installation look good?
- Can you afford both a new stove and a firesafe installation? If the installation includes a new chimney, figure on spending at least double the stove price for the installation. This will include chimney, stovepipe, fire-protective wall paneling, and protection for the floor.
- Are you willing to submit your plan to the local building inspector? This usually means paying a small fee for a permit and then another small fee for an on-site inspection, leading to a "certificate of occupancy." The inspector may first require that you revise your plan in accord with local codes. Some people regard this requirement as a violation of their rights. But it's intended to ensure sound construction and, in this case, prevent a fire. By announcing that you are putting in a new chimney, you may be letting yourself in for an increase in the assessed valuation of your home. In some localities, this may mean higher taxes. In other energy-conscious communities, the wood-burning installation may make you eligible for a tax reduction. *But either way, get the okays from officials before you start and then before you fire up the stove.*

PLANNING A NEW HOUSE. Count yourself lucky. You're in command. You can create a versatile, energy-efficient heating system that incorporates a conventional furnace as a mere backup. Solar heating ought to be in your plans, even if you feel you want to wait a few years before doing the installation. Then wood can be either your mainstay or your prime backup.

First, study some of the many excellent books available on home construction that employs solar and wood heat. Two of the key factors are home shape and superinsulation. Also important are protection from the wind, a southern exposure of windows, and unimpeded sunlight for most of the day.

You can also save money on labor and materials if you build a multi-fuel chimney with separate flues for the furnace and for each wood heater. This may also let you avoid being assessed for more than one chimney. Here just be sure that flues are separated according to the code and that flue joints are at elevations at least 7 inches apart.

Chimneys rising entirely inside a house can contribute or drain heat, depending on how you use the flues. There may be advantages in using both clay-tile flues and prefabricated insulated-metal flues. We'll cover this a bit later in this chapter.

USING AN EXISTING FLUE. Many older homes have chimneys with more than one flue. Here the builders undoubtedly sought savings in materials and labor—and perhaps taxes too. If you don't know whether or not you have an extra, just count the number of "appliances" now vented into flues, such as a furnace and fireplace, and then go up on the roof to count flues. If your count yields an extra or two, you may have a good thing, or you may not.

If the extra flue ever saw use, you'll probably be able to see traces of soot and creosote inside. If you do, you know that the old T that received the stovepipe was either filled with mortar or sealed with a metal blank cover. But you've still got some inspecting to do from the chimney top. If both the extra flue and any working flues have been protected from rain and snow by a chimney cap, the extra flue will likely be in as good condition as those in service. But if the flues haven't been protected, you should check for ice damage and general deterioration.

Lower a trouble light down each of the flues. If the working flues are coated with creosote and soot, you may have to clean them before you'll see much. For now, let's assume that they're all clean enough to inspect and that they're all straight so that you can see to the bottoms. Check the flue walls for cracked mortar and flue tiles. If it appears that there's been a lot of deterioration inside, you may remind yourself that ignorance had been bliss. For you may be looking into a honeycomb that should be condemned by a fire inspector. If things look bad inside, you'll want to perform a smoke test in each of the flues. To do this, simply build a small, smoky fire in the bottom of the flue and cover the top with a wet blanket. Then you can carefully check for escape of smoke into adjacent flues and from exterior joints. If you detect leaks, you have only two worthwhile options: to build a new chimney or to reline each flue to the satisfaction of a competent official.

We'll take up relining flues and building new chimneys shortly. But first let's get a better look at that extra flue. If it's so clean that it looks virgin, it may never have received a T connection (or thimble) for a stovepipe. In this case the builder may have figured the extra flue would come in handy but never got

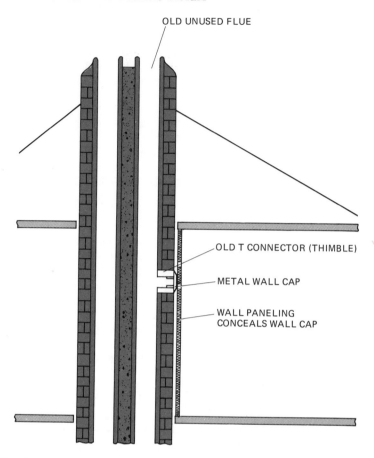

OLD UNUSED FLUE

OLD T CONNECTOR (THIMBLE)

METAL WALL CAP

WALL PANELING
CONCEALS WALL CAP

While checking the condition of an old flue, determine if a T connection was ever installed. If a T exists, it may be hidden behind wall paneling. You must find it, otherwise hot flue gases from a new installation could start a fire where the old T emerges.

around to tapping into it. Then too, a T may have been put in when the chimney was going up. Then it may have been sealed with mortar or a metal blank. If a T exists, you must find it. As you lower that trouble light, look for the shadow of an opening. If you see it, adjust your light so that it's directly in line with the opening. Then mark the light cord where it aligns with the chimney top and pull it up. You now know how far down the opening is and which wall of the chimney it's in.

Next remove any wall paneling for a good look at the blank cover or the mortar seal. If it's located right where you'd like to connect a stovepipe, you're in business. If it's not located suitably, you'll have to open it and reseal it so that the new seal provides as much fireproofing as a modern all-fuel flue does. Next you can punch an opening in the flue where it will do you the most good. And you'll sleep better knowing the old opening is sealed properly.

Is the old flue safe? Don't assume the flue is safe or that it's the right size for your stove. Again, if it shares the chimney with a working flue, chances are that it's in as good condition as its mate. After checking for deterioration with a light and after performing the smoke test, you may conclude that the flue is sound. Then again, you may not.

If the flue is an old one, it may simply be a shaft 8×8 inches or larger, with mortar or brick walls. Such a flue should be brought up to standards established by the National Fire Protection Association before you use it. Bear in mind that a flue should have a cross-sectional area at least 25 percent larger than that of the stove flue collar and stovepipe. This will help ensure adequate draft under normal conditions. Since many wood stoves in all output ranges are available with flue collars as small as 5 inches in diameter and cross-sectional areas of 15.7 square inches, you'll be able to find a quality stove, large or small, for a relined flue. Here you just have to be sure that the flue winds up with a cross-sectional area of 15.7 square inches plus 25 percent, or about 20 square inches. But building codes may decide minimum flue size for you. For my concrete-block chimney, the inspector required a clay-tile flue with a 49-square-inch cross section (7×7 inches square).

Patching an old flue. Flues are too long and narrow to allow close inspection, and you can't reach in far enough to patch them with a trowel. Faults should be patched. But at best, this patching should be considered only a preliminary step— not a complete means of rehabilitating a chimney. Here's what's involved.

A block of wood about an inch smaller in cross section than the flue is inserted in a fabric sack lined with straw filler. You'll need a round block for a round flue and a square block for a square flue. The idea is then to move the tight-fitting sack up and down in the recently cleaned flue while pouring furnace cement or fireclay cement on top of the bag. With bamboo rods connected with hose clamps, you can work the bag up and down so that it forces cement into cracks and faulty joints.

This is a good way to start patching an old flue. The trouble is, many people figure this patching results in a first-class, all-fuel, reliable flue. It doesn't. First off, the new cement often doesn't have good enough ledges in the old masonry for solid purchase. The old cracks may be lined with soot and creosote that prevent strong bonding. And in time, the cement will dry, shrink, and become worthless if relied on as the flue liner.

Relining an old flue. After patching the old flue as well as you can, you should reline it. Bear in mind that the finished interior cross section need be only 25 percent larger than that of the flue collar on your stove. And many old chimney flues are monsters, designed to handle the heavy volumes of air drafted through fireplaces and old-fashioned leaky stoves, with room for Old St. Nick. So for a stove with a small collar 5 inches in diameter, a round flue need have only a 6-inch diameter, and you could get by with a square flue as small as 4½×4½ inches.

The new liner itself should be made either of approved clay tile ⅝ inch thick, or single-wall stainless steel, or double-wall insulated stainless. These materials

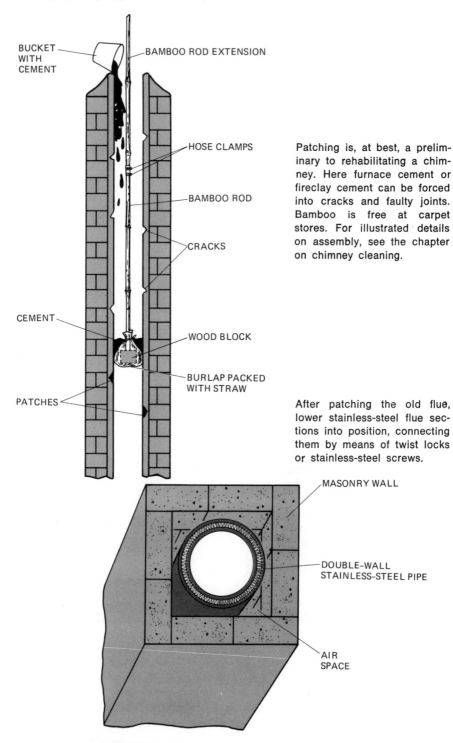

BUCKET WITH CEMENT

BAMBOO ROD EXTENSION

HOSE CLAMPS

BAMBOO ROD

CRACKS

CEMENT

WOOD BLOCK

BURLAP PACKED WITH STRAW

PATCHES

Patching is, at best, a preliminary to rehabilitating a chimney. Here furnace cement or fireclay cement can be forced into cracks and faulty joints. Bamboo is free at carpet stores. For illustrated details on assembly, see the chapter on chimney cleaning.

After patching the old flue, lower stainless-steel flue sections into position, connecting them by means of twist locks or stainless-steel screws.

MASONRY WALL

DOUBLE-WALL STAINLESS-STEEL PIPE

AIR SPACE

are known to be long-lived when assaulted by acids, moisture, and the intense heat of chimney fires. Of course, heavy-gauge stovepipe could be lowered inside with its joints held by sheet-metal screws, but within a year corrosion could become a hazard.

Some masons know how to lower clay tiles in sections with fresh mortar on the top end of each section. Ideally then, each section is bonded to the section that preceded it. But this is tricky work, calling for the right tools and spacers. It also means that you are trusting the mason to provide tight joints even though he can't inspect them closely. There's a simpler, safer way.

You can lower single-wall or double-wall stainless-steel sections secured by patented twist locks or by stainless screws. Some authorities recommend pouring noncombustible insulation between the stainless pipe and the old flue—this to promote flue temperatures hot enough to retard the condensation of creosote. But this could be dangerous. If moisture got in, it could compact the insulating material, reducing its insulating value. Also, the moisture could settle in the insulation, freeze, and expand—exerting sideward stresses on the chimney. Air space between the stainless pipe and the old flue serves better to keep the old chimney from overheating, and the air gap allows any moisture to drain to the base of the flue and out the cleanout door.

But before you attempt the relining job, get the okay from local officials on all your proposed methods and materials.

USING ONE FLUE FOR TWO OR MORE "APPLIANCES." It's common practice for heating contractors to vent a gas or oil furnace and, say, a gas water heater into the same flue. And makers of wood-only furnaces encourage the venting of their furnaces into the same flue that serves a gas or oil furnace—*under the proper conditions*. But the National Fire Protection Association advises against using more than one heater per flue.

The main problem is draft interference. Also, if one of the "heaters" is a fireplace, a wood stove using the same flue could easily send sparks into the room through the fireplace opening. Makers of wood furnaces may simply advise that the cross-sectional area of the flue be as large as the combined areas of the stovepipes it services. Here a flue area 25 percent larger would be in closer line with safe practice. Yet even though the old flue has a larger cross section than those of the proposed pipes, it may be too small.

Here the National Fire Protection Association subscribes to guidelines published by the American Society of Heating, Refrigerating and Air-Conditioning Engineers, Inc. (ASHRAE). In this case a flue's capacity is based on the Btu *inputs* of the heaters and not just on the cross sections of their vent pipes. Conventional furnaces are rated by their maximum *inputs*—the heat generated, regardless of how much reaches the home. Wood stoves and furnaces are often rated by their *outputs*—the amount of heat yielded to the home, regardless of the amount wasted up the flue. So if a wood stove is 50 percent efficient, its *input* would be twice its *output*. Here the input for a stove with an output of 50,000 Btu would be 100,000 Btu. From there it's a simple matter of adding the wood stove *input* of 100,000 Btu to that of the regular furnace, which might be 125,000 Btu. The total 225,000 Btu should then be compared to guidelines established by ASHRAE.

For example, you need a flue with a 50-square-inch cross section for a 10-foot-high flue if you want it to handle up to 175,000 Btu. A 15-foot flue with that same cross section can handle up to 210,000 Btu because its height ensures better draft. A 20-foot flue could handle up to 240,000 Btu, a 30-footer 275,000 Btu. For higher Btu inputs you need to use a flue with a cross section of 78 square inches, in which case a 10-foot flue can handle up to 300,000 Btu, a 15-footer 360,000 Btu, a 20-footer 415,000 Btu, and a 30-footer 490,000 Btu. But if you can afford to heat a home at some of the higher Btu ratings noted, you can probably also afford to provide a good, firesafe flue for each appliance and not try to cut costs by tapping into a single flue.

Also, be sure each stovepipe entering a flue enters at an elevation far enough from other pipes to minimize draft interference. Be absolutely sure no two stovepipes enter the flue at the same level.

And remember, if you use more than one pipe per flue, you are acting against recommendations of the National Fire Protection Association. By all means get approval from local officials first.

If you decide you need a new chimney, you must choose between masonry chimneys and prefabricated metal ones.

MASONRY CHIMNEYS. Masonry offers three basic choices: stone, brick, and concrete block. All should be lined either with approved clay tile or with stainless-steel pipe as described earlier in this chapter for relining an old flue. Both tile and stainless steel are capable of surviving chimney fires, and they are both reasonably immune to the effects of moisture and acids.

All masonry chimneys should have their own foundations, independent of the house foundation, and the foundation must extend below frost level. Although masonry chimneys must be anchored to an outside wall or otherwise braced when they run inside a house, these anchors and bracing are intended only to prevent tipping and not to provide vertical support.

To put up a masonry chimney along an outside wall, you'll need scaffolding. The portable metal towers are best. They can be rented or borrowed. But they are essential, because ladders just don't give you the base of support you need for handling heavy stone or block and then leveling and squaring the courses. Also, since scaffolds don't lean against the house the way ladders do, they don't damage siding.

There's something about bare stone or brick that most people find beautiful, inside a home or out. But bare gray concrete block leaves a house with an unfinished look. Painting the block improves the appearance but still leaves the "lake-cottage-economy" look. Yet blocks can be plastered to achieve any of a variety of pleasing finishes.

Because masonry is a good conductor of heat, a masonry chimney inside the house can give you 5 to 15 percent additional heat from the wood burned. So if you plan to fire your stove or wood furnace mainly around the clock, an indoor masonry chimney is the way to go. Also, this same chimney will keep flue gases hotter than a similar chimney would outdoors, and hotter flues tend to accumulate less creosote.

Since masonry is a good conductor, it is obviously a poor insulator. Outdoor masonry chimneys tend to lose heat quickly through their walls. This rapid heat loss results in flue temperatures that encourage the formation of creosote, while

Here are the basic options for masonry wall materials. Note the variety of flue caps.

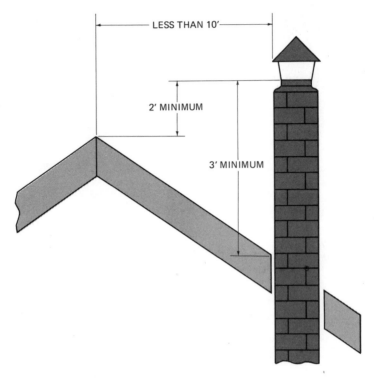

LESS THAN 10'

2' MINIMUM

3' MINIMUM

The chimney top must be at least 3 feet above the highest part of the roof it passes through. And the top must also be at least 2 feet above any portion of the roof within 10 feet, horizontally.

diminishing draft. So it's usually best to avoid building an outdoor chimney of masonry unless it is lined with an insulated flue. However, outdoor chimneys are generally less costly than indoor ones, because they need only a foundation below frostline, anchoring to the house wall, and a bit of carpentry and flashing where the chimney passes through the eave. Indoor chimneys involve either a foundation under the crawl space or a special foundation in the basement floor. Then there's carpentry at each floor and through the roof.

Stone chimneys. Of the three basic masonry types, a stone chimney will often be the most expensive if you have to buy the stone and hire the mason. But if you have an old stone fence you don't mind parting with, you'll save the bulk of the cost for materials. So stone chimneys may cost anywhere from almost nothing to several thousand dollars, depending on the availability of stone, the reputation of the mason, and the height and design of the chimney.

Brick chimneys. Brick chimneys may be as time-consuming to build as stone chimneys because of the small size of the bricks and the constant need for leveling and squaring. Brick chimneys built by first-timers are usually distinguishable

The flue tile inside these 17×17-inch blocks has a 7×7-inch interior and an 8¼×8¼-inch exterior. This flue offers a 49-square-inch interior, and suppliers refer to it as 8-inch flue tile. Leveling must be precise, both for plumb, as shown, and for horizontal level, corner to corner. Note the insulating air space between the flue and the block; this must not be filled with mortar.

Heavy blocks require the strength and control of two stout men working from solid scaffolding. Don't attempt to handle such large blocks alone or from only a ladder.

BASICS OF BLOCK CONSTRUCTION

Clay tile is brittle. You can cut the top section as shown by marking the outside and filling the inside with sand as high as the marked level. Then the chisel is less apt to crack tile below sand level. Some masons use a hacksaw for cutting.

Whether finishing block, brick, or stone, you'll want to improve the weatherseal and appearance of joints by means of a joiner. Use a trowel to catch excess mortar.

from far off. Be concerned about fire safety as well as looks, too. Unless you have experience laying brick, a brick chimney may be a big challenge. Here you'd probably be better off helping a mason or even splitting firewood. You can get reliable cost estimates by asking several contractors to make bids. They can tell you what materials to order and then work on a day rate themselves.

For 8-inch-long chimney bricks, you'll need about seven bricks per square foot of surface on the chimney. If bricks cost you 15¢ apiece, you can figure that a 17×17-inch chimney 24 feet high will cost $143 for bricks alone. Flue tile for that chimney will probably cost another $40. Mortar and sand for the bricks may run another $50. Concrete may run $10 to $20 for the foundation. A cast-iron cleanout door, anchors, and flashing may run another $15. So materials for a 24-foot brick chimney would probably cost about $260. A good mason could probably build such a chimney up an outside wall in a couple of days.

Concrete-block chimneys. These are popular because of economies in materials and labor. A skilled mason normally figures on running a 30-footer up an outside wall in a single day. So if you opt to tackle the job yourself, you'll probably be saving only a day's cost in labor, while adding at least a couple of days to your own labors. The 17×17-inch blocks in my own chimney weigh almost 50 pounds apiece. That's not terribly heavy at ground level, but lifting the blocks to roof level and working with them overhead is work for two stout men. So unless you're experienced, the boss man might as well be a real pro or else someone's brother-in-law who's built sound chimneys before.

Blocks with 17×17-inch exteriors will house what's called an 8×8 flue tile. This tile has a 7×7-inch interior and about an 8¼×8¼-inch exterior. With mortared joints, it takes thirty-five 10-inch-high blocks for a 24-foot chimney (the same height as the brick chimney we costed above). And flue tile will cost the same as for the brick chimney, about $40. The thirty-five blocks at $2.50 each will cost about $88. Mortar and sand will cost a little less than for a brick chimney because there are fewer joints. Figure four bags of mortar at $3 each and eight bags of sand at $1.75 each for a combined total of $26. The foundation and other hardware will cost about the same as for the brick chimney—say, $25 to $35. That totals to about $180 for materials for a 24-foot concrete-block chimney.

So, generally, materials for a brick chimney will probably cost over 40 percent more than for a concrete chimney, and labor time will be about double. It's reasonable to figure that a brick chimney will cost about twice as much as a block chimney of comparable size.

PREFABRICATED METAL CHIMNEYS. These are double- or triple-wall insulated pipes of various convenient lengths. Most feature a simple twist-lock connection that makes them easy to install. Don't use the low-temperature insulated pipes designed for gas or oil heaters. You need the all-fuel type approved for wood heating by the Underwriters Laboratories.

Ideally, both walls are corrosion-resistant stainless steel. Insulation between chimney walls should be effective for high temperatures. The outer walls may be painted to mask the chimney's vacuum-bottle-liner appearance.

Installation instructions and safe clearance vary, depending on design and insulation ratings. Provided you are using UL listed parts, just follow manufacturer instructions.

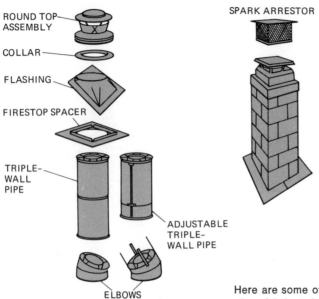

ROUND TOP ASSEMBLY

COLLAR

FLASHING

FIRESTOP SPACER

TRIPLE-WALL PIPE

SPARK ARRESTOR

ADJUSTABLE TRIPLE-WALL PIPE

ELBOWS

Here are some of the basic components of prefabricated chimneys.

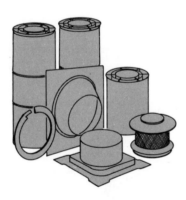

Prefab advantages. Prefabricated metal chimneys enjoy wide popularity over masonry chimneys for several reasons.

The materials cost is generally as much as for a brick chimney, but installation is easy enough in most cases for the do-it-yourselfer. The dealer may make a low-fee installation for you.

Since these chimneys seldom weigh more than a few hundred pounds, they don't need costly masonry foundations. Instead they can be mounted in the ceiling directly over the stove, or they can be bracket-mounted to an inside or outside wall at some point higher than the stove or furnace. This results in a shorter overall length than for a comparable masonry chimney that must have a foundation below frostline.

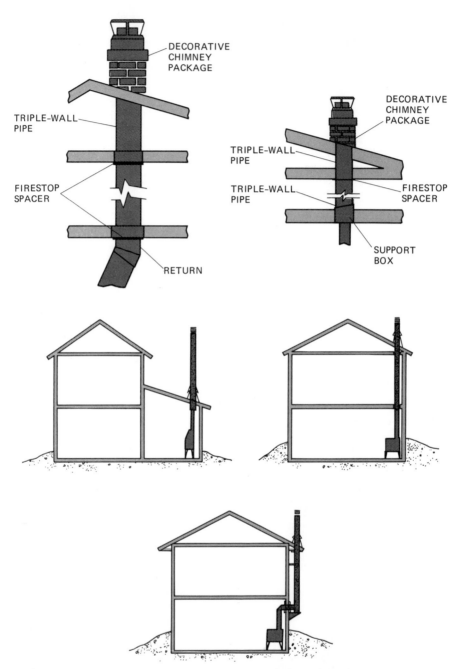

Prefabricated metal chimneys offer a host of do-it-yourself options. These provide just a sampling. Avoid bends and elbows if possible because they tend to collect creosote, and they make cleaning more difficult. *Note:* All pipe passing through combustibles must be protected by fittings bearing UL approval.

Prefab chimneys also enjoy the endorsement of the National Fire Protection Association. Here the chimney's insulation results in higher flue temperatures than occur in comparable masonry chimneys—especially those running up outside walls. Again, this reduces accumulation of creosote.

They are also superior to masonry chimneys when used only intermittently, because they heat up faster. The cold mass of masonry robs flue heat and causes heavy condensation when the stove is started up and until the masonry is warmed sufficiently.

Prefab drawback. My own chief objection to prefab chimneys is their appearance. Bare metal inside a house doesn't harmonize with most decors, and running up the outside, it looks like a hastily chosen afterthought. These chimneys can be painted, though, and there are imitation brick shells that run the range from convincing to chintzy. Or you can build masonry around them. These measures, of course, drive up the costs for materials and labor considerably.

CHOOSING BETWEEN MASONRY AND PREFAB CHIMNEYS. Like most choices, this one involves compromise and establishing personal priorities. If you plan to heat around the clock and want a chimney that will transfer heat to the room through which it rises, then a tile-lined masonry chimney will serve better. You'll increase stove efficiency 5 to 15 percent. But if you fire only intermittently, that same chimney will absorb room heat through its walls to the cold air inside the flue. So for intermittent firing an interior chimney that is insulated will rob less room air when the flue is cold, and it won't take as long to warm up at start-up time. This also helps prevent condensation of creosote at startup. Yet for steady, around-the-clock firing, an insulated chimney is still hot enough to deliver heat to the house. In fact, its shell may get hot enough to burn flesh. The exterior of an insulated metal chimney is merely less hot than it would be if it were a single-wall metal pipe. So an insulated chimney for a stove fired around the clock may add as much as 5 percent efficiency to the stove's output.

If aesthetics, space restrictions, or installation costs make an interior chimney impractical, you can always run the chimney up the outside of the house. To minimize creosote formation resulting from cool flue temperatures, you should use a prefab metal double-insulated chimney. The sections allow a low-cost installation, but they also are the least attractive option. You can run a double-wall steel chimney inside a masonry chimney for better looks and less creosote formation, but at higher cost.

My own chimney is concrete block, lined with tile. It runs up one side of the house 24 feet. It does collect creosote, but I use a homemade cleaning brush that makes the flue easy to clean—almost enjoyable. The cooler flue condenses tars and volatiles that would escape into the outside air were my flue insulated. So I don't feel bad knowing the flue is uninsulated. It's helping reduce a little air pollution. As long as I tend to my cleaning chores, the creosote won't be much of a fire hazard. Besides, clay flue tile is far less expensive than stainless steel. As I write this, flue tile costs only about $1.50 per foot. Insulated stainless steel costs $12 per foot.

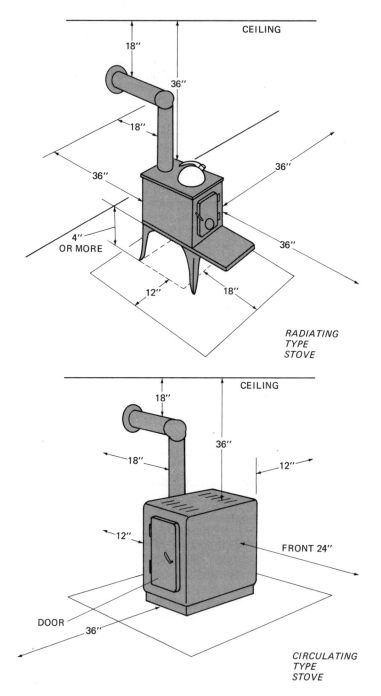

These drawings show minimum allowable clearances from *unprotected* surfaces, as established by the National Fire Protection Association. In this case, even noncombustible plaster, gypsum, and asbestos are considered unprotected surfaces unless spaced from other combustibles because they can conduct intense heat to wood supports.

CLEARANCES. The National Fire Protection Association has established minimum recommended stove clearances, because heat from stoves can ignite nearby surfaces. Minimum clearances for radiating-type stoves are more stringent than those for circulating stoves. This is because radiating stoves heat nearby surfaces more. Circulating stoves heat primarily by convection, with buoyant air rising between the firebox and the stove jacket. The radiant energy emitted from the jacket is far less intense than that from a radiating stove.

You can reduce the recommended 36-inch clearance for radiating stoves, the 12-inch clearance for circulating stoves, and the 18-inch clearance for stovepipes only when you use specific combinations of protective asbestos, metal, fiber insulation, and air space. These are listed and illustrated in the accompanying drawings.

It's important to note that asbestos and plaster—and even fireproof gypsum— may require just as much clearance as wood paneling. This is partly because enough heat can be conducted right through them to ignite wood support strips, and because nails in these noncombustible covers can heat up enough to char and ignite wood on the other side.

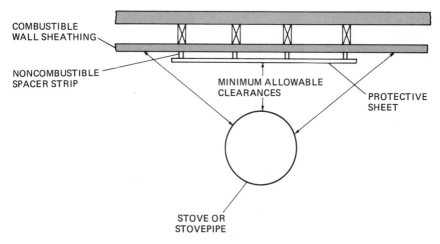

This drawing and table, adapted from National Fire Protection Association guidelines, shows the distances to which minimum clearances may be safely reduced when the correct combinations of air space and noncombustible sheeting are used.

TYPE OF PROTECTION	WHEN UNPROTECTED SURFACES ARE					
	36"		18"		12"	
	ABOVE	SIDES & REAR	ABOVE	SIDES & REAR	ABOVE	SIDES & REAR
¼"-inch asbestos spaced out 1 inch*	30	18	15	9	9	6
28-gauge sheet metal spaced out 1 inch*	18	12	9	6	6	4

* *Noncombustible spacers*

Floor clearances are comparatively lenient because the ash, sand, or firebrick in stoves helps insulate stove bottoms. Yet radiant energy can ignite a wooden floor just as fallen embers can. You should provide at least 4 inches of clearance, and 10 inches is safer and reasonable. Then you should cover a wooden floor with mortared bricks or stone, or with 24-gauge sheet metal over ¼ inch of asbestos. It's important to mortar the bricks rather than just set them in place. This will retard transfer of radiant energy, and it will also keep fallen embers from getting between the cracks as you tend the fire. The NFPA doesn't provide much guidance for floor protection. But you probably should extend it at least 12 inches out from the sides of the stove and at least 18 inches in front of the door.

Never let the metal legs of a stove touch a combustible floor, even if the legs are long and provide lots of clearance. The stove legs may transfer enough heat to ignite any combustible material they rest on.

STOVEPIPES AND CONNECTING HARDWARE. Fortunately, most stove makers and retail outlets offer pipe and connecting hardware suitable for their stoves. The National Fire Protection Association provides these guidelines:
- The pipe should be at least as large as the flue collar on the stove.
- For pipes of 10-inch diameter and less, the steel should be at least 24-gauge. (Remember, the lower the gauge number, the heavier the steel.)
- Short, straight runs are recommended because they result in less cooling of flue gases. This helps maintain adequate draft and minimizes condensation of the tars and acids that are in creosote. Short runs of pipe also mean fewer pipe joints that could fail, and shorter runs of pipe are easier to clean than longer ones. In my own house, I initially experimented with a pipe configuration that ran 11 feet and made three sharp 90-degree turns before joining the chimney flue. My intent was to increase stove efficiency by promoting maximum heat transfer through the pipe. Here stove efficiencies can be increased anywhere from 10 to 15 percent. But since I was using a highly efficient stove anyway, temperatures inside the pipe were low. These low flue-gas temperatures combined with the draft impedence of the three elbows and the long horizontal runs in the pipe to build up creosote and ash, especially on the stove side of each elbow. This "creoash" also lined the walls of the pipe, top and bottom, and served to insulate the pipe, minimizing heat transfer to the room. So my elbows and long pipe were really causing buildup of insulating ash that prevented heat transfer. Lately, I've done as well using a shorter pipe with only one 45-degree and one 90-degree elbow. Provided you use an airtight, efficient stove in the first place, you're well advised to keep the total run of pipe under 6 feet. All horizontal sections should pitch upward at least ¼ inch per foot and should never be more than 75 percent of the length of the chimney above the top of the pipe.
- Never let stovepipe pass through other than a masonry wall or ceiling unless you use a vented or insulated thimble listed by the Underwriters Laboratories. In this case, follow the manufacturers' specifications to the letter. Generally, ventilated thimbles have a diameter at least 12 inches greater than the stovepipe, or metal or fireclay thimbles must be surrounded by brick or fireproof material with a diameter at least 16 inches greater than that of the stovepipe. Or you can cut a hole in a partition wall, leaving at least 18 inches of space on all sides of the stovepipe; the opening should be sealed with sheet metal.

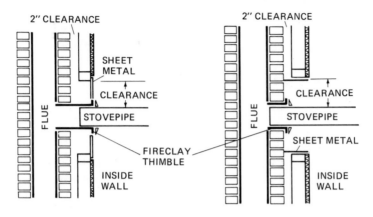

Here are two methods of running stovepipe through a combustible wall. Clearance options are noted in the text. (Drawings are adapted from a National Fire Protection Association publication.)

• Provided you have a good chimney, a good stove, and proper stove clearances, the stovepipe and its connectors provide about the only other opportunity to make an unsafe installation. Many home fires result every year because of well-intentioned shortcuts here. If the maker of your stove or the stove retailers indicate you can make installations less fire-wary than I've outlined here, go with my recommendations. Mine follow conservative but fire-wise guidelines of the National Fire Protection Association. These are designed to protect you under the worst possible eventuality—the stovepipe and chimney fire. But if your stove maker or retailer recommends more stringent precautions than I've outlined, go with the more stringent ones. Design features of the stove may warrant the more stringent precautions.

• Then maintain your wood-burning system as described in Chapter 9.

9 | How to Clean Chimneys and Stovepipes

EFFICIENT, AIRTIGHT WOOD stoves and furnaces tend to have cooler flue gases than fireplaces and leaky wood burners because they transfer most of their heat to the house. And when burning at slow rates, such as overnight, the efficient wood burners allow increased percentages of volatiles and tars to escape unburned. Then, with relatively cool flue temperatures, volatiles condense on the walls of stovepipes and flues—forming creosote.

All wood-burning devices, including fireplaces, tend to deposit the highest amounts of creosote when they're just starting up. At those times, hot flue gases assault a cold flue, and this brings condensation on flue walls. Also, at startup time many people use resinous conifer wood because it ignites at lower temperatures than most hardwoods. Resins condense as a sticky creosote.

Chimneys themselves can encourage creosote formation. Masonry chimneys on outside walls lose heat through their own walls. Metal insulated chimneys and masonry chimneys that rise inside a house often have flue temperatures as much as 200°F warmer than those of outside masonry chimneys, and with all other factors equal, chimneys with warmer flues collect less creosote—but all chimneys collect some.

There's just about no way to avoid some creosote formation. Even though the best-burning hardwoods lack the resins that guarantee creosote buildup, they produce flue temperatures that condense creosote if they have a high moisture content or if they are banked for low, even-burning overnight fires.

Some chimneys may not show heavy buildups for many years, but most chimneys giving steady service should be cleaned at the end of each heating season—if not to prevent chimney fires, then at least to remove the acidic deposits that can gradually corrode mortar and stovepipe. And all chimneys should be checked at least before each heating season and then about every four cords of wood thereafter.

CHIMNEYSWEEPS. The easiest but most expensive way to clean a chimney is to hire a chimneysweep. The profession got a boost with the energy crisis of the early 1970s. Modern-day sweeps don't look much like the frail, wheezing and coughing lads of the past who usually descended right into the chimneys. There, amid a storm of soot, ash, and creosote, the lads would brush and scrape until they were given a breather—the chimney a little cleaner, their lungs a little blacker.

Today's sweeps may arrive with a truckload of equipment that includes respirators, industrial vacuum cleaners, an assortment of large steel brushes, fiberglass extension rods, extension ladders, ropes, tire chains, burlap bags, and safety belts.

Some of the less energetic sweeps never go up on the roof at all. Instead, they place the hose nozzle of their 700-cubic-feet-per-minute vacuum right in the maw of the fireplace or at the cleanout door of a chimney for a stove. With the vacuum on, they shove the flue brushes up by means of the extension rods. The soot and creosote fall into the ravenous hose. And that's that. But that kind of job compares to visiting a dental office to have your teeth cleaned without letting the dentist check for cavities.

Better chimneysweeps, those who may also serve on the local fire department, give your chimney a careful inspection with a sweep. For that they must go up on the roof as well. To them, the profession is something more than one with "low overhead, flexible hours, low initial investment, and high visibility," as one dealer of cleaning equipment promises.

So instead of phoning only the sweep with the lowest advertised price, you should phone a few and quiz them on their methods. If no sweeps advertise in your area, your local fire inspector may be able to recommend a few sweeps who know their stuff—doing the preventive maintenance that helps keep the fire engines at the station.

CLEANING YOUR OWN CHIMNEY. Aside from the dollar savings, there are advantages to cleaning your chimney yourself. First, it encourages your giving the chimney a careful inspection. And second, by cleaning yourself, you know how thorough the cleaning job was.

The importance of thoroughness was illustrated to me by Carl Bruder, a neighbor who custom-makes fireplace dampers. One day, as Carl tells it, he was called to a grand old home to take measurements for a damper. When Carl arrived, the owner lamented that the offset (bent) flue had been cleaned only a week before, but that it still didn't have the draft it once had.

Carl sent his helper onto the roof and had him lower a tire chain. As the chain disappeared into the offset in the flue, it encountered an obstruction. The helper drew the chains up a bit, and then dropped them. No success. Carl then told the helper to pull up the rope and add more chain. The load now weighed over 40 pounds. The helper dropped it repeatedly against the obstruction. Suddenly, the heavy chains broke through, and an avalanche of branches from a squirrel's nest clattered down onto the hearth, accompanied by a distressed squirrel that left its tracks all over the house before it escaped. Yes, there's merit in doing a job oneself.

HOW TO CLEAN A CHIMNEY. First, check to see if the flue needs cleaning. This should be done at least at the end of each heating season so you have time for repairs should you need them. Then just before the heating season, you should make a quick check for squirrel or bird nests. Bird nests will have no tenants by this time, and you'll be saving the squirrel from a warm and gassy demise. Then the flue should be checked again during the heating season after it has handled the smoke of four to six cords of wood.

If you have a fireplace, remove andirons, ash, and damper. Then shine a bright light up into the flue. Probe the flue with your finger and a putty knife. If creosote is evident there, it will surely occur farther up the flue where flue gases have cooled and condensed even more creosote. If the flue and smoke shelf look pretty clean and if you see mostly bare flue all the way up, you can probably postpone cleaning. If you see a lot of soot and creosote, prepare to clean. This can be done best from the roof. So seal up the fireplace opening, using the regular glass doors if you have them and either a panel of wood or a tight-woven drop cloth. This will keep falling debris and ash from entering the room.

If you have a wood stove or furnace, the first check should be the cleanout compartment at the base of the chimney. If it contains a mound of fallen creosote looking like charred cornflakes and chunks of prunes, you can count on some creosote higher in the flue. These fragments in the cleanout compartment result from drying and cracking creosote higher up. If the cleanout door contains only a little ash, your flue may be relatively clean. If you also find bits of mortar and flue tile in the compartment, plan to locate the source.

Whether you have a fireplace, stove, or wood furnace, the best vantage for cleaning is usually from the chimney top. When you go up, wear soft-soled shoes with a good grip. If the roof is so steep that walking is risky, you can lay a ladder on the roof with a safety bracket hooked over the roof peak. Most chimney tops will be low enough so that you can look into the flue while standing on the roof. If the chimney top is so tall that you can't peer inside without using a platform or ladder, pause to reconsider. A platform or stepladder can damage the roof cover unless it is set on a protective base. Then too you should secure the platform or ladder to the chimney itself, making your stand absolutely solid and immovable. If the ladder leans against the chimney, be sure the chimney won't topple under the stresses you will cause while ramming and pulling contraptions through the flue. Here it's prudent to wear a safety belt and secure it to some anchor on the other side of the house. Drainpipes are not reliable anchors. Nor is the family car, unless you carry the only set of keys. A mature tree is a good anchor. For a safety line, use quality ¾-inch manila or ⅜-inch braided-nylon rope, and protect the roof peak where the line passes over. Then tie the line to yourself so short and taut that it allows you to see into the flue but not lean farther than necessary. That way if the chimney topples, you'll drop straight onto the roof and no farther.

If the chimney cap is the removable type, remove it. If it's permanently bonded to the chimney, you may be able to poke your head inside for a look or you may have to use a mirror. A permanent cap, such as a slab of slate mortared to piers of corner bricks, limits your view and also restricts working room.

If your chimney is insulated stainless steel running up the outside of the house, you may want to partially disassemble it for a cleaning at ground level, or you may opt to scrub it out the same way masonry chimneys are.

Masonry chimneys generally have a beveled mortar seal that slopes from the flue down to the chimney edge. This helps drain rain and condensation from flue gases to the outside of the chimney. Check the seal between the flue tile and the mortar. Any crack there or elsewhere in the mortar should be patched to prevent water from entering, freezing, and thereby expanding and forcing more cracks or breaking out sections of flue tile. Such cracks tend to compound themselves, each one allowing moisture to reach new places.

Also check for cracks along exterior joints and where the flashing makes a water seal between chimney and roof. To check the effectiveness of the flashing

An inexpensive chimney cap can be removed after screws at its base are loosened. Note bevels on the mortar that promote rain runoff. Check for, and seal, any cracks between the flue and the mortar.

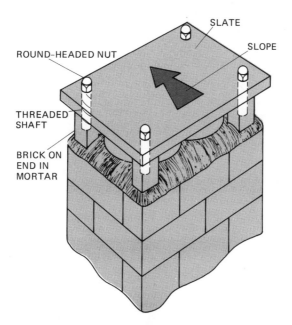

Here is one means of making the popular slate chimney cap removable for flue cleaning. To help runoff, the slate should be sloped slightly toward the low side of the roof.

Inspect the chimney top for cracks, which sometimes result from ground settling, sometimes from ice damage. Old mortar can be chipped out and replaced.

around a chimney that passes through the roof, you can look for watermarks from inside the attic. If the chimney passes up an outside wall, faulty flashing can allow water to dam up between the house and chimney. Damage here can be catastrophic, because it's not noticeable until an ice-damaged chimney has heated the house walls to the point of charring them and causing smoke.

If you suspect that cracks may have caused further damage or if damage looks severe, proceed with cleaning anyway. Then you and the fire inspector can get a better look at conditions inside the flue. Chimney repairs are described in Chapter 8.

Next inspect the inside of the flue. If the flue is sooty, the beam from even the brightest flashlight will be absorbed within 5 or 10 feet from the flue top. But since the top of the flue tends to collect a fair share of the creosote anyway, its condition is a good indicator. Scrape through any soot with a putty knife or steel brush. Creosote here may be thin and flaky as well as thick and rock-hard—like black peanut brittle.

If the top few feet look pretty clean, with plenty of bare flue showing, things probably won't look any worse farther down. But it's good to check anyway. You can do this by lowering a powerful flashlight on a line or by lowering an AC trouble light by its cord. If things still look clean, check for corrosion in a steel flue or for cracks and rotten mortar in a masonry flue.

If your flue is offset, with a crook in it that prevents seeing to the bottom, remember that crooks and bends impede smooth flow of flue gases enough to cause eddies that deposit ash and creosote. (For his home in Hartford, Connecticut, Mark Twain once designed a fireplace with two flues branching around a window that was set over the hearth so that he could watch the fire and look out the window at the same time.) Anyway, if the flue takes a sharp bend, you may have to lower tire chains or a burlap bag full of tire chains to see what they knock down or how much soot the burlap picks up.

Many do-it-yourselfers like to use tire chains on the end of a rope, or in a burlap sack. Others prefer to weight a sack full of chicken wire or branches, sometimes tying a rope at each end and using two people to pull from above and below. In the old days, people commonly pulled small evergreen trees through the flue. Less often, they used a protesting goose.

Many professional sweeps like the chain method best. Others, and some homeowners, prefer to use large steel brushes that are shaped a lot like household bottle brushes. These may be cylindrical for flue walls or square for flue corners. The brushes can be lowered by rope with a 15-to-20-pound weight hanging from the bottom, or they can be connected to a succession of flexible rod extensions and

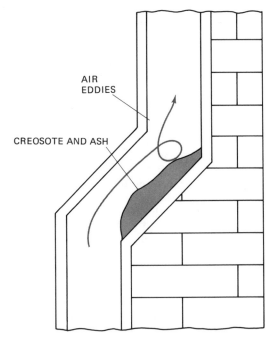

Offset flues tend to accumulate excessive amounts of creosote and ash on the lower portion of the offset. This can choke down the flue enough to impede draft greatly. And heavy accumulations become dangerous stores of fuel for a chimney fire.

Steel flue brushes may cost over $15 a piece. The X-patterned bristles help reach corners of square flues. The loop on top of each attaches to a pull rope. The bottom loop is fitted with a 20-pound weight.

shoved through the flue from the top or the bottom. Then, as noted earlier, large industrial vacuums are employed to catch falling particles or to clean up after. There are also bladed contraptions that can be lowered for scraping. Sticky creosote resulting from the burning of resinous woods is the toughest stuff to remove. But for a do-it-yourself tool, you can straighten the neck on a garden hoe and fasten it to bamboo extension rods, as I'll describe shortly.

Though all of these methods work well for some people, I would caution against using the chain as a battering ram, forcing it through brittle flue tiles if it gets stuck. The result can be cracked tiles and loosened mortar. Then, too, don't let chains strike a fireplace damper that may be brittle after years of use. If you can borrow an industrial-type vacuum, great. It will pull in a lot of fly ash as well as heavy stuff. But its filters should be designed to trap sub-micron particles. Woven filters may let the particles right through and into your house. Under no circumstances should you use your household vacuum. It will blast fine dust into the room and probably into motor bearings. As an alternative to a vacuum, just seal off the fireplace opening or plug the connection between the stovepipe and the flue. Then you can simply wait for the dust to settle and go in with a shovel, and later a broom.

If you have a straight flue no more than 40 feet long, you can rig up your own brush-and-pole contraption that cleans on the principle of two toothbrushes taped back to back and run through a narrow tube. I came up with the design after finding a simple straight steel brush very effective in cleaning stovepipes. To make this cleaner, I fastened two steel brushes back to back on a spacer block. I fastened bamboo rods to the brushes and block. The bamboo extensions can be secured with hose clamps. The contraption cleans very well indeed. It's lightweight, sturdy, and easy to use. Hardware, complete, should cost less than $5. Details on construction and use are shown on accompanying pages.

Note: If you suspect leaks in your chimney, you can test for them by building a smoky fire and plugging the chimney top with a wet towel.

HOW TO CLEAN A STOVEPIPE. Stovepipes should be cleaned more often than chimneys. Generally, the shorter the stovepipe, the hotter its gases as they enter the chimney flue. Any factor that promotes heat transfer from stovepipe walls into the room causes a cooling of the stovepipe gases. Many people attempt

(Text continued on page 100.)

DO-IT-YOURSELF CHIMNEY BRUSH

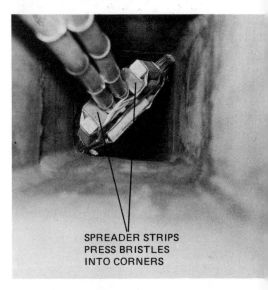

SPREADER STRIPS
PRESS BRISTLES
INTO CORNERS

This flue was once fouled with heavy soot and creosote. Note the added spacer strips used to broaden the head enough to reach the diagonal corners. For cleaning the flat flue walls, a 2-inch-wide set of brushes would give better control than the 1-inchers shown because they can better resist the twisting of the head as it is jammed up and down.

Here the head of a homemade chimney brush starts down the flue, applying sufficient pressure against flue walls to remove all ash and baked-on creosote. The bamboo poles, free from carpet stores, allow excellent control of the cleaning head. Best effects are earned by an up-and-down jamming motion.

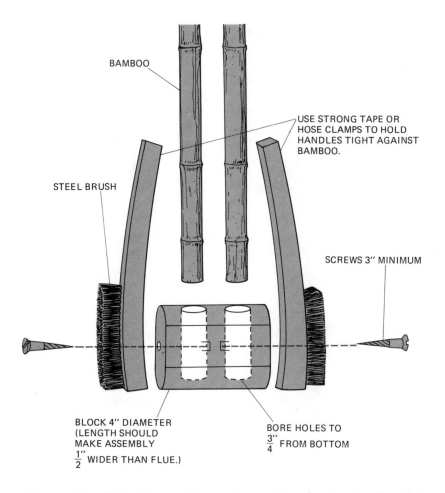

BAMBOO

USE STRONG TAPE OR
HOSE CLAMPS TO HOLD
HANDLES TIGHT AGAINST
BAMBOO.

STEEL BRUSH

SCREWS 3" MINIMUM

BLOCK 4" DIAMETER
(LENGTH SHOULD
MAKE ASSEMBLY
$\frac{1}{2}$" WIDER THAN FLUE.)

BORE HOLES TO
$\frac{3}{4}$" FROM BOTTOM

This assembly will handle round flues and opposite walls of square flues, but it won't be broad enough to reach diagonal corners. So the screws should be long enough to allow insertion of spreader strips, shown in previous photo, between the block and the brushes.

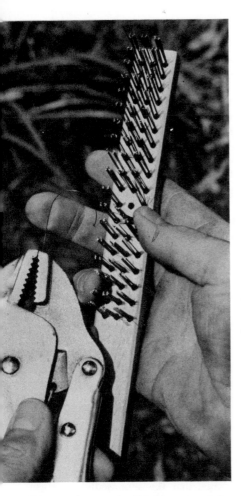

Prepare the way for the long screws by first removing wires from a hole near the center of the brush. Then drill through.

Bamboo extensions are best secured with hose clamps.

to gain a 10 to 15 percent increase in heat transfer, and thus an increase in wood-fuel efficiency, by employing long stovepipes, or stovepipes with several elbows that impede air flow. Some people attach metal fins or doughnuts to the stovepipes to extract heat. Some send the flue gases through smoke chambers between the stove and the chimney. And some coil water pipes around the stovepipe to heat the water in a hot-water tank. All of these options can improve heat transfer. But if the stove is efficient in the first place, one often winds up cooling stovepipe gases to levels that result in increased condensation and creosoting, as well as a reduced draft that doesn't carry smoke along with much force.

The result of maximum heat transfer through a stovepipe is often a buildup of creosote, ash, and a mixture of the two that I call creoash. This poses a fire hazard certainly, for if it catches fire it can heat your stovepipe cherry-red and maybe generate thunderous drafts that stress pipe joints until they separate. As well, these deposits become effective insulators after a time. They may eventually insulate so well that the stovepipe transfers very little heat to the room. So they defeat the purpose of long pipes and other extractors by causing insulating buildups. In addition, the acids in creosote are murder on steel pipes. Cheaper, thinner pipes can corrode beyond safe use in a single heating season, and the buildups can choke the pipe diameter enough at elbows and along horizontal runs of pipe to further impede the draft.

So stovepipes deserve periodic checks throughout the heating season. One of the simplest checks is to snap a fingernail against a straight section of pipe. If you hear a ping, that portion is relatively free of ash and creosote. If you hear a dull thud, you can figure on some creosote or heavy ash. Horizontal pipe sections and bottoms of elbows tend to collect the most gunk. So it's best to use only one or two elbows with gentle bends and to minimize the amount of horizontal pipe.

The next step is to wait until the fire has died in the stove and remove any warm ash outdoors. Next you can spread a protective cover over the floor to catch pipe ash and then begin disconnecting the most suspect elbow. If the elbow is fouled, you should clean the whole stovepipe. If you failed to remove all ash from the stove, you may notice with annoyance that the supposedly dead ash in the stove has begun to give off smoke. To remedy this, just reconnect the pipes and take all ash outdoors.

If feasible, take the pipe outdoors without disconnecting sections. But before you head out, plug the hole leading to the chimney. If you don't, the natural flue draft will rob the house of heat all the while the pipe is disconnected. And by plugging the hole, you will prevent unruly flue drafts from dumping flue ash into the room. Also, if you expect that you may have to separate a pipe of many sections into three or more sections for cleaning, it's wise to apply a piece of masking tape to each individual section and number it in its assembled sequence. Some of the mildest people have been brought to rage over mixups during reassembly.

To clean the pipe, you may have to scrub hard with a wire brush. If your arm isn't long enough to reach all the way to an elbow, you'll have to disconnect another section. If you have trouble pulling sections apart, be patient and resourceful. Sometimes they are fused by creosote or corrosion. Too much force can bend pipes and split lengthwise seams. Usually you can succeed by plant-

This creosote and ash buildup occurred at the third of three 90-degree elbows in 9 feet of pipe. Horizontal runs and elbows tend to collect the most "creoash." Pipe joints should also be secured by three screws each.

Steel brushes let you scrape clean enough to check for corrosion. The brush shown is one of those shown earlier in the flue-brush assembly, and here it's doing double-duty.

ing one foot in the opening of an elbow and administering a kind of pulling bear hug on the section you want to remove. Then by alternating bear hugs with hand slaps on the selected joint, you should be able to work the section free. That failing, use a lubricant such as Liquid Wrench.

If you are performing these maneuvers on a cold winter day, you'll want to carry the cleaned pipe sections inside before attempting reassembly. Pipes get darned cold on a cold day, and mittens may be a hindrance.

Reassembly time may be the occasion you vow to shorten a long stovepipe. For even though you match up all the joints correctly, the mating sections may refuse to mate. It's best to resist clobbering them together, for if you do, you may only split a seam. The problem may be an end that is out of round, or corroded, or still fouled with creosote. After you've checked for those possibilities without finding the problem, take the male section back outdoors so it gets cold again and contracts. When you bring it in cold, it should slip home. If not, try a few more drops of lubricant. Then, with the stovepipe sections reassembled, set the unit aside.

Now it's time to clean out the stove. Ash and creosote can insulate stove parts so that they absorb and transfer less heat than they should. Give the interior a good brushing down. If there are removable baffles, take them outside for cleaning.

Now clean out the flue thimble in the wall. That's it. The cleaning job is done. You can reinstall the stovepipe.

Since the reinstalled stovepipe may drip some creosote during the first hours of operation, it's wise to leave the floor cloth in place directly under the pipe. Soon the liquid creosote will bake within the joints enough to form a seal. Then you can remove the cloth and enjoy the stove during the weeks or months until the pipe no longer pings when you snap it. Then it's time to clean again.

The pencil points to a ¼-inch ash buildup on top of a stove baffle. Since ash insulates, stove walls and baffles should be cleaned regularly in order to promote the best heat transfer to the room. You should also check that the gasket around the door is in place and making good contact with the stove's door lip.

10 | House-fire Preparedness

A WOOD-BURNING SYSTEM is as firesafe as any fossil-fueled heating system if it is well designed and well made and if it is installed, operated, and maintained properly. Yet there's something about heating with wood that engenders a new concern for fire safety. Perhaps it's the loading of fuel wood. Perhaps it's striking the match. Perhaps it's just the blast of heat and light you encounter whenever you gaze into the fire chamber. No longer is fire an automatic comfort controlled by unseen valves, pumps, and the innards of thermostats. Now it's an extension of yourself—under your control or beyond it, as circumstances have it. Thought naturally turns to fire detectors, fire extinguishers, and fire drills.

FIRE DETECTION. Most residential fire victims die from smoke and toxic gases rather than from burns. These fires, often merely smoldering ones, take heavy tolls at night, when people are asleep. In most cases, residential fires produce detectable smoke before they generate enough heat to activate commercial heat alarms. For these reasons, experts recommend smoke detectors over heat detectors, when you can afford only a unit or two initially. Yet heat detectors can also play an important role in your fire-warning system.

Most smoke detectors are engineered to detect small amounts of visible smoke. Some also detect invisible smoke during a fire's initial stages. Smoke detectors are sensitive enough to sound the alarm when there's a heavy concentration of tobacco smoke or excessive downdrafting from the chimney flue. The idea is to announce real fire in time to allow everyone to leave before being overcome.

Heat detectors activate when the air around them reaches a predetermined temperature or heats up rapidly. For most rooms that don't normally reach temperatures of 100°F, fixed-temperature detectors are set to activate at temperatures between 135°F and 165°F. For areas such as attics and furnace rooms that can be expected to have temperatures of over 100°F, the detectors are set at 175°F or sometimes higher.

Required standards for smoke and heat detectors. Whether battery- or AC-powered, all devices must have enough power to sound an alarm for at least four minutes.

AC devices for do-it-yourself installation should not exceed 30 volts, and they must have a visible light that shows that power is on. They should not be plugged into an outlet that is controlled by a switch, and the plug should be fastened so that it can't become dislodged.

Battery-powered devices must have a battery with an expected minimum life of one year, and this includes weekly power tests. These units must have an audible trouble signal that will sound at least once per minute for seven days, should the battery weaken or the terminal contacts corrode.

Here are components of the First Alert smoke detector by Pittway Corporation. This unit can detect both visible and invisible products of combustion. It employs a tiny, harmless, radioactive source that produces electrically charged molecules (ions) in the air. This sets up a current within the detector chamber that is reduced when smoke particles attach themselves to the ions. The reduced flow sets off the alarm. Other detectors may employ a light bulb inside a darkened chamber. When smoke particles enter, they reflect the bulb's light into a light-sensitive cell, and this sets off the alarm.

Combination fire and burglar alarms should emit different sounds for each type of hazard.

Be sure the unit is listed by a nationally recognized testing authority.

Power source. As a primary source of power, the National Fire Protection Association recommends commercial AC over batteries, but batteries have its approval. When batteries are used as backup sources, they must be capable of sustaining the system for twenty-four hours and then sound the alarm. Some heat detectors are nonelectric, relying on mechanical responses and discharges from small compressed-gas tanks.

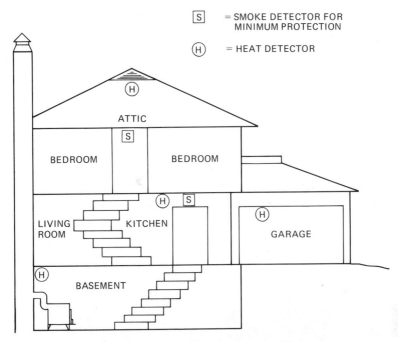

A multi-level house might be equipped with detection devices as shown. Priority should go to the smoke detectors at the tops of stairwells. Bedroom protectors are of special value if they contain a person who smokes. A heat detector rather than a smoke detector is used in the kitchen because cooking smoke might set off unnecessary alarms. Since some smoke detectors may not function when temperatures are below 40°F and above 100°F, heat detectors may be warranted in attics and garages—and over furnaces and stoves. If soundproofing could muffle alarm sounds from any area, the alarms should be linked to a central alarm.

Installation. Smoke detectors should be where you expect smoke to pass, such as along the top of a stairwell, and also near bedroom entryways, where detection is vital. It's smart to have at least one detector on every floor. Follow manufacturer's instructions.

The popular "spot-type" heat detectors carry a rating for their effective "square of protection." That is, installed in the center of a ceiling, these devices will sense excessive heat for an effective distance of, say, 10 feet, or whatever the unit is rated for. Thus, it pays to locate such a unit its effective distance from any wall in order to get maximum coverage from the sensor. Avoid mounting these detectors within 6 inches of the joint made by wall and ceiling. Dead air in this region retards heat penetration and so make heat detectors ineffective. Also, if there are open joists in the ceiling, heat detectors should be installed on the bottom sides of joists; this promotes most rapid sensing.

FIRE EXTINGUISHERS. Commercial extinguishers for the home may have one or all the following classifications:

Class A: for ordinary combustibles, such as wood, fabric, and many plastics
Class B: for flammable liquids, gases, and greases
Class C: for electrical fires, when it's important the extinguishing substance not conduct electricity

The National Fire Protection Association (NFPA) advises that if you have only one portable extinguisher, it probably ought to be a multi-purpose dry-chemical type that is classed for A, B, and C type fires. Be sure the device is listed by a national testing laboratory, though.

Popular stored-pressure models weigh from 3 to 20 pounds themselves, with capacities generally about half the weight of the unit. These portables have discharge times of only 8 to 25 seconds. So don't count on the smallest ones to smother a raging holocaust. Most are activated by means of a combination lift-

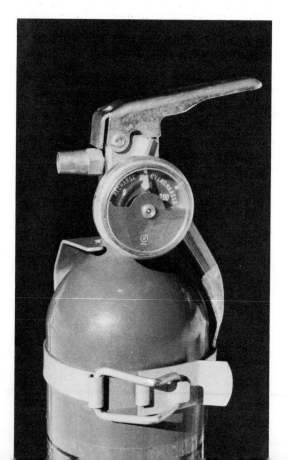

This shows the squeeze handles and nozzle on a small pressurized dry-chemical extinguisher. Note that the pressure gauge indicates a 100 percent charge. If pressure drops, the needle will indicate recharge.

trigger and squeeze-handle. The discharge should be aimed at the base of the flames, not just into the smoke.

These pressurized extinguishers have a pressure gauge that tells you whether the unit is operable or whether it needs a recharge. Be sure to check this gauge at least monthly. If the pressure gauge shows a low reading, take the extinguisher to a commercial recharge station listed in the Yellow Pages under "Fire Extinguishers."

It's smart to have more than one extinguisher. Here the NFPA says a good second choice would be a 2½-gallon water type that is activated by either stored pressure or by a manual water pump. But remember, this is for a Class A fire, normally associated with burning wood—not for volatiles or electrical fires.

Portable extinguishers loaded with Halon are fairly new. Halon is a liquefied gas that is blended with nitrogen. The mix breaks down a fire's fuel molecules. Shot at high pressure, the mix allows you to stand farther from the blaze than standard dry-chemical models do. Some Halon units are classed only for B and C fires. Others carry the full A-B-C classification.

Storage of portable extinguishers. Extinguishers should be stored in the normal path of escape from places where fires would be most apt to originate—such as near chimneys, stoves, furnaces, and the kitchen range. This way, if you attempt to extinguish a fire and fail, you can still escape.

Automatic sprinklers. These are a system of water pipes running to sprinkler heads along the ceiling. Most heads are designed to emit spray when a soldered link in them is melted by temperatures of about 165°F. Have a licensed plumber make the installation.

Water hose. A simple garden hose stored indoors offers an unlimited discharge time. So if you exhaust the supply from your portable extinguishers, a water hose can help until the firemen arrive. The hose should be hung, at the ready, and should be long enough to reach any part of the house—interior or exterior. If convenient, during the heating season, keep the hose connected to an indoor faucet. Do not leave this hose connected to an outdoor faucet during cold weather, for it may collect undrained water and then freeze. Even a dry hose left outdoors may be too stiff to manipulate when cold. And don't depend on connecting the indoor hose to an outdoor faucet, which may be frozen fast when you need it.

Water buckets. A 5-gallon water bucket, with lid and padded seat cushion, can double as stool and extinguisher. The rig can be handy near a fireplace or stove for splashing out renegade ash and embers, or very small fires. A bucket of sand is also useful.

ESCAPE PLANS. Since most fatal house fires occur at night, your escape plan should focus primarily on night alerts, yet it should include daytime maneuvers too.

Each bedroom should offer at least two exits—either two doors or a door and a window. The window should be easy to open for the occupants in all seasons. That failing, each occupant should know how to break through glass and screening with a chair, and then knock glass shards from the frame before cushioning the frame edge with a blanket and crawling out. Upper-story rooms should offer some means of descent—whether by stairs, rope ladder, or fireman's pole.

Small children should sleep near enough to the parents for a quick rescue. Here an adjoining bedroom door is an advantage. If the family is large, with bedrooms scattered about the house, there should be a prearranged meeting place outside. That way firemen will know which rooms may still contain someone. For this contingency, fire departments often issue decals that can be affixed to windows of bedrooms containing invalids or small tots. Unfortunately, these decals may give information to burglars too. There's a risk either way.

Periodic fire drills should include simulated escapes. Older children and adults should be taught how to administer mouth-to-mouth resuscitation so they can help anyone overcome by smoke. Everyone should be taught that a smoke-filled room usually offers a stratum of reasonably breathable air about 18 inches off the floor, right at crawling level.

The National Fire Protection Association advises sleeping with bedroom doors shut so that smoke and flame can't enter bedrooms as easily. Of course, this serves both to protect you from asphyxiation and from the rapid spread of fire throughout the house. But this preventive has drawbacks that you'll have to weigh in light of your own situation. It may prevent hearing a smoke or heat alarm, or smelling smoke until the fire has progressed mightily, or hearing cries for help from other bedrooms.

11 | Comforts, Cooking, Fire Tending

THERE'S NO HOT water like water heated on a wood stove. Begin with a whistling tea kettle with at least a half-gallon capacity. The best and safest style is dome-shaped, its only opening a spout covered by a hinged lid that can be raised by squeezing a spring-loaded trigger. This is the safest style because it won't let much scalding water escape if you should stumble with it. Kettles with open spouts and lift-off lids are okay as long as you never walk more than a few steps with them. But you'll probably find, as we did, that stove-heated water is made for walking.

The tea kettle in the photo is about the safest design If you carry its boiling water around the house. Note the multi-hinge tongs. For stoves with small ash aprons, place a coal scuttle underneath each time you open the door.

First, there's boiling water for tea and other hot drinks. There's always plenty of water for the dishes. And you've never taken a bath until you've taken one in hot water from a big tea kettle. On a cold winter's night, after a day of wood splitting, there's nothing quite like tea-kettle-prepared bath water, reheated after you've soaked awhile by yet another dose from the kettle. Of course, this indulgence requires that you have a partner to bring water. Then you can return the service when it's her turn.

If low room humidity becomes a problem, the kettle can sit on the stove to humidify your house. A boiling kettle can add many gallons of moisture to the air during a day. This will help you keep relative humidity in the 30 to 50 percent range. Otherwise you may wind up with a dry nose, squeaky furniture, static electricity, and parched house plants. Adequate humidity makes you feel warmer than in dry air of the same temperature because there's less cooling evaporation from your skin. Note: Avoid over-humidifying. Too much humidity can cause condensation on house sheathing and structural members.

If the whistle on the spout is too shrill for comfort, you can alter it. If the whistle consists of a simple hole in the lid, bore the hole a little larger. If the whistle is a mechanical contraption, just fiddle with it until it produces a more soothing sound. When our kettle is relatively full, the hiss of its steam is counterpointed by jolly glub-glub sounds of boiling. When the kettle is nearly empty, its hiss is accompanied by thin and anxious bubbling. So without lifting the kettle to test its weight, we know about how much water remains.

SIMPLE COOKING. The amount of cooking you do will depend on the stove's location and your yen for small amounts of extra work. In our home, we avoid all frying and sloppy cooking because our stove is in our ground-level family room. Wood cook stoves work well, but they should be in a kitchen where you have access to vents, sink, and the dining table.

Aside from cleaning chores, cooking should also occasion thoughts on safety. Soups and stews can cause serious burns, should you spill them or fall with them. Wheeled food carts can reduce risks. Stairs are prime hazard areas. If you must carry food upstairs, use pots with lids that can be clamped shut. You can limit your stairway hauling to foods such as rice, with liquid boiled off, or else you can pour off hot water from vegetables before you truck the vegetables upstairs.

The great difference between cooking on a wood stove and on a conventional range is that you must adjust the placement and elevation of the cookware rather than merely adjusting the heat output. If your stove has a large top, you'll be able to locate areas of higher and lower heat. Practice will tell you where they are. Yet, even then, spots of lowest heat may still be too hot for your needs. In this case, you can elevate the cookware by placing it on a cast-iron trivet or on any of several coil trivets you can fashion from door springs and coat-hanger wires. A large-diameter spring may give you simmer to medium, depending on where you place it. A small-diameter spring may give you medium to high heat. Practice, with different fires, will make you the world's expert on your own stove. Then it's a matter of checking the pots occasionally for desired effects.

NO-SWEAT BAKING. The ovens of wood cook stoves are gems. If you have such a stove, count yourself well provided for. But if you don't have a built-in oven, you can rig a stove-top model for about $1. Or you can buy a stove-top

Here's the $1 foil-tub oven that can be used for everything from warming a slice of pizza to baking bread. Bakeware should be elevated on a trivet to prevent scorching the bottom of the food. Here an old cookie rack serves as the trivet.

oven, complete with temperature gauge, for about $40. We've found the $1 version adequate for all baking from breads and biscuits to potatoes and yams.

The idea is to create a hot-air chamber around the food, making sure that baking pans don't rest directly on the stove. For this, simply place the baking pan on a trivet of some sort and then cover the works with an inverted aluminum-foil tub. Such tubs are available in most supermarkets. They dent easily, but with careful use and storage, they'll last indefinitely.

We baked our first experimental loaf of bread too fast in our foil-tub oven—ten minutes—but aside from a hard bottom crust, the bread was good. Once you become intimate with your stove, you'll be able to guess pretty accurately whether you've got the 350°F you want for cakes and biscuits or the 400°F you want for breads and pies. Of course, you can use a thermometer, but that's the cautious way.

Baking inside the stove. This is campfire cookery indoors. Basically, we're talking aluminum-foil cookery—foil wrapped around the food and laid in coals. Potatoes done this way bake in ten to fifteen minutes. Although there's almost no limit to the dishes you can prepare this way, the foil becomes brittle and unusable after a time or two. Stove-top baking works as well without wasting foil.

DEHYDRATING FOODS. People dehydrate foods to preserve them and to reduce their weight for backpacking. All you need are a stove, drying racks, and airtight containers for storage. There's almost no limit to the foods that can be dried: fruits, vegetables, meats, herbs, seeds. Drying saves most of the nutrients.

The idea is to evaporate moisture from the foods at temperatures ranging from 105° to about 140°F. Methods vary depending on whether you use an enclosed stove-top dehydrator or a simple sandwich grill, which doesn't retain moisture as an enclosed unit does.

FIRE-TENDING HARDWARE. You will need several items to handle burning fuel, clean up the hearth, and store wood.

Ash scuttle and shovel. The scuttle can serve both to carry ash outdoors and to catch embers that tumble out when you open the stove door and tend the fire. The shovel has many uses, including ash removal, log manipulation, and ember heaping. Its handle should be long enough to let you reach to the back of the stove without burning your hands.

Multi-hinge tongs. These give you good gripping power and extend your reach over 2 feet. The multiple hinges allow you to take a wide bite without opening the handles wide. This lets you reach deep into a stove with a small door in order to manipulate the logs and embers with eyebrow-plucking precision. Since the tool lets you position logs exactly where you want them, it can help you achieve highly efficient burns. A good set of wrought-iron tongs costs about $12.

Whisk broom. You'll need a small broom to sweep up ash and wood debris near the stove's loading door and near the wood-storage rack. Never use your home vacuum on the ash, because the fine particles will pass through the dust bag, entering the room and fouling motor bearings. Besides, if a vacuum picks up embers, you could wind up with a fire in the dust bag.

Ash rake. For stoves with a long, horizontal fire chamber, an ash rake helps you pull embers forward in a heap prior to laying the next load of logs. Simple hoelike devices homemade from sheet metal are adequate, or you can bolt a metal plate to the threaded end of a metal rod. For shorter stoves, a shovel or set of tongs works well enough.

Fireplace irons. These normally include a long-handled shovel, a hooked poker, and a long-handled broom. Some sets include a pair of tongs shaped like a tuning fork. But these tongs don't offer enough leverage for a powerful bite, even though they hang more pleasingly from a stand than multi-hinged tongs do. Prices for irons on a stand begin at about $100. For fireplaces, they are okay, but for stoves you'll do better with the tools described in preceding paragraphs.

Indoor wood rack/carriers. Because of the possibility of insect invasions, it's best not to store more than a half-day's supply of wood indoors. That way you allow the wood to achieve room temperature without fully awakening dormant creepers. So you don't really need a large indoor rack. The handiest racks serve both as rack and carrier. The best designs let you carry the wood suitcase-fashion so that you can maintain good balance on icy walkways or porches. They should be simple, open frames that let you place the wood close enough to the stove to evaporate any surface ice or moisture.

The Sheep Rancher Cook Stove by the U.S. Stove Company has a sheet-steel body and a cast-iron top. A duplex draft system allows rapid heating of the oven, which measures 9×13 inches in front and 16 inches deep. The stove stands 29 inches tall and weighs 105 pounds. It is shipped assembled except for the legs, which require four bolts each.

The Victor Junior Wood/Coal Range comes from Pioneer Lamps and Stoves in Seattle. This unit comes with optional nickel plating, a hot-water reservoir (on the right side), and a warming oven (the part above the cooking surface).

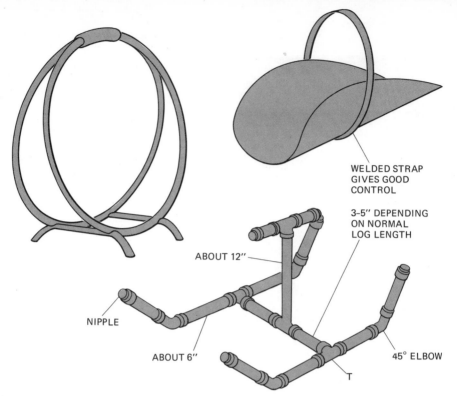

WELDED STRAP
GIVES GOOD
CONTROL

3-5″ DEPENDING
ON NORMAL
LOG LENGTH

ABOUT 12″

NIPPLE

ABOUT 6″

45° ELBOW

T

Here are three practical wood racks that also serve as carriers. They let you carry the wood low, keeping your center of gravity low as you walk on icy porches or walkways. The basket style and the welded-rings style generally cost over $20. You can fashion the ½-inch-pipe style in a hardware store or plumbing shop and then let the clerk count up the bill, in amazement. Figure on about $10 for pipe.

This wood carrier was originally a crate for milk bottles. Made of galvanized steel rods, it is sturdy enough to carry about 60 pounds of wood. Placed at the business end of a sawbuck, it catches poles and small logs for easy carting and keeps them out of the snow. It also serves as a container for tools and accessories.

12 | Wood Fuel

ONE GOOD WAY to remember a series of facts is to put them into verse. Cadence and rhyme fix the items that might otherwise elude memory. In this regard, an anonymous English poet has provided a delightful three-stanza ditty on firewoods, and the work has found its way into some of the best American wood-heat literature. But as you enjoy the ditty, remind yourself that there are many types of beeches, birches, oaks, and ashes. In this case, an Englishman's tree of a given name may not be the one you've come to know well. Let's consider the stanzas one at a time.

> Beechwood fires are bright and clear
> If the logs are kept a year.
> Chestnut only good, they say,
> If for long 'tis laid away.
> But ash new or ash old
> Is fit for queen with crown of gold.

The American beech, ranging in the East, does not grow well when planted in England. Its dense wood weighs 40 pounds per cubic foot when air-dried to a 12 percent moisture content. The air-dry weight is important, because it tells you there's a mass of wood with comparatively high-density material available for combustion. Freshly cut, or "green," beech will have roughly a 63 percent moisture content. Yet this is a moderately low water content among green hardwoods. Even so, as the poet implies, a year of aging (under well-ventilated cover) reduces a wood's moisture content to 10 to 20 percent. This brings the wood nearer its burning potential. (Note: Percent moisture content should be thought of roughly like proofs in alcoholic beverages. For example, liquor that is 100 proof is only half alcohol. In this case, the proof is a ratio of alcohol to water, rather than a ratio of alcohol to the whole volume of liquor. Wood scientists figure moisture content much the way distillers figure alcohol content. Basically, they set up a ratio, finding percent moisture content by dividing the weight of water in the wood by the weight of the wood when oven-dry. The resulting percent figure should be thought of somewhat like proof in liquor and not as a percent of the whole mass of wet wood.)

Unfortunately, fungal blight of the American chestnut has about eradicated this once important eastern tree. Sprouts still rise annually from the stumps of dead or felled chestnuts, but they normally survive only a short time before they are killed by the blight, which is harbored in the stumps. If you find a large healthy specimen, don't cut it down! Instead, contact a forester, for you may have discovered an immune individual that will bring botanists in droves. Besides, this chestnut was never renowned as a powerful heating wood. It is one of the lighter hardwoods, weighing 27 pounds per cubic foot when air dry.

Of the dozen or so types of ash in America, only five are abundant, and four of these are basically eastern or midwestern trees. Queens would like our white ash, whether "new or old," because it has only about 45 percent moisture content when green. This is about the lowest moisture level of all the major hardwoods, and the result is that freshly cut ash burns fairly well, though it burns better when seasoned. Its 37½ pounds per cubic foot after drying makes it a medium-heavy wood that provides good heat relative to its volume.

> Birch and fir logs burn too fast,
> Blaze up bright and do not last.
> It is by the Irish said
> Hawthorn bakes the sweetest bread.
> Elm wood burns like churchyard mold,
> E'en the very flames are cold.
> But ash green or ash brown
> Is fit for queen with golden crown.

Yes, the paper birch, sometimes called the canoe birch, burns fairly rapidly. It weighs about 34½ pounds per cubic foot when air-dry. Its 80 percent moisture content when green and its watertight bark dictate that you split the logs immediately so that moisture has a chance to evaporate. Unsplit logs may rot in one year. Sweet birch (often called black birch) and yellow birch (with peeling gray bark) are heavier birches, weighing 38½ and 40½ pounds per cubic foot, respectively, when air dry. They provide good to excellent heat value relative to volume, but their wavy grain makes them fairly tough to split.

Like all lightweight woods, the firs burn rapidly. Air-dry, they weigh only 20 to 25 pounds per cubic foot. Like most conifers, or softwoods, firs contain resins that condense in a stickier, pitchier creosote than results from burning most hardwoods.

Of the 800 or so species of hawthorn in America, about 150 grow large enough to be considered trees. They look so much alike they are hard to tell apart. The "haw" here means "fence" or "hedge," and the "thorn" growing at the base of the twigs and leaves keeps grazing animals away. These small trees serve very well as hedges. And, as the Irishman implies, the wood is hard and heavy—ideal for the long, steady burn with high heat output needed for a day of bread baking.

American elms throughout North America are dead or dying from a fungus that causes Dutch elm disease. The fungus is introduced into the tree by elm bark beetles when they feed in twig crotches. Development of the fungus in the twigs and branches causes wilting and eventual death. The beetles breed under the bark of dying or dead elms or elm that has been stacked for firewood. The fungus fruits under the bark, and emerging beetles then carry the fungus to other elms. So any elm cut and stored for firewood should first be stripped of its bark to prevent breeding of the beetles and to prevent eggs from hatching or young from developing. Anyway, city ordinances may prohibit the collection of elm for firewood. So that's worth a check.

The finger points to a larva of the European elm bark beetle, which carries the Dutch elm disease. The vertical and radiating galleries were caused by these larvae, although the galleries themselves are not symptoms of the disease and usually occur in elms that are already dead or dying.

It's true that American elm is not a powerful source of Btu. It's of medium weight among hardwoods, weighing about 31 pounds per cubic foot when air-dry. Elm's cross grain makes it the devil to split, and the fibers contain a relatively high 94 percent moisture when green. Once split and seasoned, though, this elm provides a moderately good heating fire. It leaves more ash than many other hardwoods, but the ash is excellent fertilizer. Does elm really burn like churchyard mold? Churchyard mold is a fuel I haven't tried, but poorly seasoned elm makes a damp and dismal fire.

> Poplar gives a bitter smoke,
> Fills your eyes and makes you choke.
> Apple wood will scent your room
> With an incense like perfume.
> Oaken logs, if dry and old,
> Keep away the winter's cold.
> But ash wet or ash dry
> A king shall warm his slippers by.

Broadly speaking, poplars of North America include the willows, cottonwoods, and aspens. All are lightweight, weak hardwoods. Great drinkers of ground moisture, they are usually more than 100 percent (or half) water when green. Unseasoned, they burn poorly and smokily. Seasoned, they burn up quickly, so as fuel they are valued mainly as kindling.

Apple wood, like nearly all fruit and nut woods, is heavy and burns with a hot, aromatic, and long-lasting fire. Most people burn only prunings. It's shameful to down a wild apple tree for firewood, because the tree is a valued food source for many kinds of wildlife.

When air-dry, oaks in the red-oak group (pointy leaves) are heavy, ranging from 37 to 43 pounds per cubic foot. Oaks in the white-oak group (with smooth edges or else rounded lobes on the leaves) weigh from 39 to 55 pounds per cubic foot. Only the hickory group among large trees rivals the white oaks for heat value—hickories weighing about 45 pounds per cubic foot when air-dry. The prize goes to live oaks, weighing in at 55 pounds; they hardly float. Oaks contain 70 to 80 percent water when green, and they require a full year's seasoning before approaching their heating potential. The longest-burning and best-heating wood I've ever burned was that of a heart-rotted white oak that had seasoned dry on the stump.

If you were a king with firewood savvy but without time or inclination for cutting and splitting, you'd probably direct your servants to deliver you seasoned hickories, oaks, and fruit woods. But if you were the king's woodcutter, with more chores than you could handle, you'd be wise to go for white ash and the straightest and tallest hard maples. They'd satisfy a king's demand for heat even though they are far easier to cut, split, and season than the heaviest woods. Hard maples need more seasoning than ash for a good burn, but they weigh roughly the same. Sugar maple weighs a tad more than ash and so delivers a bit more heat for the same volume. There are lighter ashes and maples too that serve fairly well.

While the king is warming his slippers, let's take a closer look at all the woods we can warm our own slippers by.

HARDWOOD AND SOFTWOOD. Botanists group trees into two basic types, based on seed structures. But for most of us, seeds may not be in season or within reach. We rely on more convenient characteristics, though they are less foolproof.

The U.S. Forest Service goes with the terms "hardwood" and "softwood." Hardwoods are nearly all deciduous, meaning they lose their leaves in autumn. Their leaves are generally broad, and the trees usually produce flowers that may range from modest little clusters to gorgeous extravaganzas. The term "hardwood" arises because most of these trees have wood that weighs over 25 pounds per cubic foot when air dry. Increased weight means increased cell density and increased hardness. These same trees are also called deciduous, or broadleaf, or flowering. All these terms are appropriate, though none is perfect, for each also describes a few trees in the other basic category—softwoods.

Softwoods are the trees that most people think of as evergreens. Yet softwoods such as the bald cypress, larches, and kinkgo lose their leaves in autumn. All except the ginkgo have leaves that are needlelike (such as pines) or scale-like (such as cedars). Most have wood that weighs less than 25 pounds per cubic foot, air-dry. This lighter wood tends to be softer than most woods of the hardwood group. Thus the term "softwood."

HEAT VALUES OF HARDWOODS AND SOFTWOODS. Using oven-dry woods, scientists have determined that nearly all hardwoods yield a maximum of 8600 Btu per pound when burned in a bomb calorimeter that allows no heat loss. The most resinous softwoods yield a maximum of 9150 Btu per pound. The higher Btu yield from softwoods is due mainly to resins and a higher lignin content.

Then why not favor softwoods for fuel? First, their unburned resins tend to condense in flues in a sticky, pitchy mass that combines with other condensing volatiles and ash to form a creosote that is difficult to remove. Then if the flue temperature rises to only 670°F it can ignite the resins, which in turn burn hot enough to easily satisfy the 1100°F ignition temperature for most other condensed volatiles. All this fuels a 2000-3000°F chimney fire. You'll find more on the disadvantages of chimney fires in earlier chapters.

Softwoods are made of large lightweight cells that burn faster than the tighter and heavier cells of hardwood. This means that stoves fired with softwoods must be loaded more often, and more logs are handled in the process. The advantage to softwoods is their ease of lighting and then their hot burn. For these reasons many people like to use them for kindling and then switch to hardwoods for sustained burns.

Still, some areas of North America offer mainly softwoods for burning, so it's often necessary to use what's available. By keeping a close vigil on creosote buildups—and cleaning accordingly—you can burn softwoods for decades with never a mishap. Vigilance!

HOW TO DETERMINE ANY WOOD'S HEATING VALUE. Aside from resin and lignin content, all hardwoods and all softwoods are quite similar chemically. It's essentially because of their resins and higher lignin content that most softwoods have the edge over hardwoods on Btu per pound.

Among hardwoods, a pound of oven-dry lightweight aspen has about the same heating potential as a pound of oven-dry hickory. Pound for pound, hardwoods of equal moisture content will yield about the same amount of heat. This is true of softwoods with equal moisture and resin contents too.

Here's the catch: An oven-dry pound of aspen takes up almost twice as much space as a pound of hickory. You'll need nearly two truckloads of aspen to match the weight of a truckload of hickory. So if you are ever given a choice of wood in a free truckload, pick the heavier wood. That way you receive more pounds of wood, and wood yields energy by the pound.

Though oven-dry weight per cubic foot is the chief determinant of a wood's burning qualities, moisture content in air-dry wood is also significant. Some heat is always lost in driving off water vapor, and this heat loss is directly related to the weight of the water vaporized.

This water has two origins. First, there's free water in the wood cells and bound water in the wood fibers. Second, there's water created in combustion when free hydrogen atoms combine with oxygen to form H_2O. Surprisingly, this newly created water will weigh over half a pound (.55 pound) for every pound of bone-dry wood consumed. That figure remains constant no matter what kind of wood you burn.

Now to the free and bound water! That's the water in the tree when you cut it down as well as any water the wood absorbs while it's stacked. This may be rainwater, ground moisture, or moisture from the air. You can measure the moisture content in any wood by weighing a small kindling-size strip that includes both heartwood and sapwood. Then dehydrate the wood by placing it in a wood-stove dehydrator or in a conventional oven overnight at 103°F. This will leave the wood oven-dry—bone-dry. Then weigh the wood again. The difference between the two weights is the weight of the water driven off. Then to determine what percentage of the sample was water, you simply divide the weight of water by the oven-dry weight:

$$\text{Percent water} = \frac{\text{weight of water}}{\text{oven-dry weight}}$$

Using this simple procedure, scientists have developed convenient tables indicating probable moisture contents of many woods. For our purposes, it's enough to know that about 1050 Btu are consumed in vaporizing each pound of water. So the drier a wood, the greater its potential heating value.

WOOD WEIGHTS AND VOLUMES. The available heat in wood depends on its moisture content in relation to the wood's oven-dry weight. Fine. The problem is, it's impractical to buy and sell wood by the pound. It's a bother to weigh wood in the first place, and if you bought by weight, you'd want to know the moisture content so that you could deduct the weight of water from the load. Freshly cut white ash may have only a 45 percent moisture content. American sycamore may have about 120 percent, and some pines may have over 150 percent. Well-seasoned woods of any type may have moisture contents of only 10 to 20 percent, depending on local humidity levels. Generally, if the ends of logs are split and withered-looking, you know the ends at least are fairly dry. Even so, center sections of these same logs could be quite moist and rotting. Whacked together, green logs emit a dull thud. Seasoned logs crack like bowling pins. To be sure of average moisture content, you'd need to dehydrate a small sample in an oven and weigh it as explained earlier in this chapter. Yet, most piles of wood contain a variety of wood types, which begin with different moisture levels and then air-dry at different rates. For sure, it's not practical to buy wood by weight.

So nearly all firewood is sold by volume. The standard measure is a full cord, which is a neatly stacked tier, or tiers, of wood measuring 4×4×8 feet. But there are cords and cords. Some suppliers call any pile of wood a cord as long as two of its dimensions measure 4×8 feet, no matter what the third dimension. These smaller cords are usually called face cords or fireplace cords. Here the lengths of logs may be anywhere from 1 foot to 3 feet, the third dimension accompanying the 4×8-foot "face."

It's easy to see that a face cord of 1-footers equals only about one-fourth of a standard cord, and 3-footers would equal only about three-fourths of a cord.

Since heating values are lower for these smaller cords, price should rightly be reduced accordingly.

Although the 4×4×8 dimensions of a cord multiply to give a potential 128 cubic feet, you won't get nearly that much solid wood per cord. A stack of round 4-foot logs, varying in diameter from 3 to 8 inches, will give you only about 80 cubic feet of solid wood. The rest is air space between the logs. Some of the air space results from the nonmating, round exteriors of the logs, but much of it results from bumps and bends of the log surfaces. The effect of bumps and bends is significant. To illustrate, if that same cord of 3-to-8-inch logs of 4-foot length were sawn up to a 1-foot length and restacked, you'd find that the shorter logs would occupy much less space than the 4-footers. Disregarding losses of sawdust in cutting, a tightly stacked cord of 1-foot logs will give you about 95 cubic feet of solid wood, rather than the 80 cubic feet that 4-footers would give. The reason is that bumps and bends don't have as great a separating effect on the shorter logs.

BUYING BY THE CORD. Here you should be chiefly concerned with the following: (1) cubic feet of solid wood, (2) moisture content, (3) types of wood, and (4) cost.

Cubic feet. We just covered the importance of tight stacking in relation to solid wood content. You should also check to be sure you are getting a full measure. In this case you should measure the wood stacked, either where you buy it or in the truck before it is dumped. Of course, you can have the stuff stacked in your yard by the supplier, but he'll bill you extra for that. If you stack the wood yourself after delivery, you'll probably have to settle any credits for shortages over the phone.

Moisture content. Oven-dry wood of a given type offers more available heat than an equal volume of green wood of the same type. At about 10 to 12 percent moisture content, a hardwood loses about 8 percent of its heating value. Each increase of 10 percent moisture lowers available Btu by about 1.2 percent. Yet a wood with 100 percent moisture content (meaning it is half water by weight) still costs you only about 13 percent of the wood's Btu value if that same wood were air-dry at 12 percent moisture. That doesn't sound too bad. The trouble is, if you try to burn only very wet green woods, you may not be able to maintain high enough temperatures within the wood itself to boil off water and generate gases for 1100°F ignition. Also, you may have to maintain a high rate of draft to keep the wood burning—robbing warm room air and consuming the wood faster than you'd need to if it were drier.

If you must burn woods with moisture contents of over 50 percent (that is, more than ¼ water by weight), it's best to burn them along with dry woods. The heat from the dry woods will drive the water out of the green woods and provide the necessary temperatures for combustion of gases from the green wood. Some people like to mix green and dry woods before bedtime. This way the green logs take time to dry out and don't begin to sustain their own flames right away. Then these green logs tend to be the last ones consumed overnight.

If you must burn woods green, try to choose woods with low moisture contents, such as white ash (45 percent). Or since heartwood is usually much drier than sapwood, you can separate the two when splitting and then burn mainly heartwood at first.

Types of wood. There are about 800 tree species in the United States and Canada, including all native species and those introduced and now reproducing on their own. Of the 800, the U.S. Forest Service has listed 109 as commercially important, based on abundance and demand. Since abundance is a prime factor in availability of firewood, the Forest Service list becomes a handy reference. I've reproduced that list in the table "How Fuel Woods Compare," appearing later in this chapter. For ease of reading, I've converted Forest Service figures for specific gravities of wood at 12 percent moisture to pounds per cubic foot, and I've entered Forest Service figures for moisture in green (freshly cut) heartwood and sapwood.

In the table you'll also find an approximate Btu value for a cord of each type of wood, at 12 percent moisture content. These are based on the high-heat value of a pound of any wood at 12 percent moisture. High heat per pound is determined by dividing 1 + moisture content into the 8600 Btu we know a pound of oven-dry wood will yield when there is no heat loss. The idea is to exclude free and bound water from consideration as a source of heat. The calculation looks like this:

$$\frac{8600 \text{ Btu}}{1.12} = 7678.6 \text{ (which we'll round off to 7679)}$$

It's also possible to calculate further losses required to vaporize the free and bound water, as well as the .55 pound of water created in burning each pound of wood. This would give a low-heat value for that original pound of wood close to 7000 Btu. Further deductions can be made for excess air in the stove, heat lost up the flue, and so on. But that gets complicated and depends on conditions of the burn in a particular installation.

In the table, I've used the high-heat value of 7679 Btu per pound of wood at 12 percent. High heat is practical here because it simplifies comparisons of the high-heat values of wood, fuel oil, and natural gas. If you know high-heat values, it's simply a matter of assigning assumed efficiencies. For example, the best wood-burning stoves and furnaces can operate at 50 percent efficiency with the draft open and at 70 percent efficiency with the draft nearly closed. You can then choose an efficiency level based on the way you burn. Oil burners and natural-gas burners operate at efficiencies ranging from 40 to 80 percent, depending on tuning, mostly. Under the best of conditions you can figure on 70 to 80 percent efficiencies from fossil-fuel burners.

The table lists 109 woods. You'll find aspen, basswood, butternut, cottonwood, and willow weighing only 21 to 23½ pounds per cubic foot at an air-dry 12 percent moisture content. These same woods have a high-heat value between 12½ and 14½ million Btu per cord. That's figuring 80 cubic feet per cord, 21 to 23½ pounds per cubic foot, and a high-heat value of 7679 Btu per pound of wood at 12 percent.

$$80 \times 21 \times 7679 = 12\frac{1}{2} \text{ million Btu}$$
$$80 \times 23\frac{1}{2} \times 7679 = 14\frac{1}{2} \text{ million Btu}$$

Since 1 gallon of No. 2 fuel oil yields a high-heat value of 140,000 Btu, all we have to do is divide the 140,000 into the 12½ million and the 14½ million to see that a cord of those lightweight woods is about the equivalent of 89 to 104 gallons of oil. To figure natural-gas equivalents, we can figure that 1 cubic foot of natural gas yields 1000 Btu. Again, it's a simple matter of dividing the 1000 into the 12½ and 14½ million Btu the lightweight woods can yield. The results show that those lightweight woods are about the equivalent of 12,500 to 14,500 cubic feet of natural gas. If you know the price you have to pay for oil or gas, cost comparisons are easy.

Use the guide below when consulting the large table "How Fuel Woods Compare," later in this chapter.

HEAT PER CORD OF WOODS (BY WEIGHT) AND FOSSIL-FUEL EQUIVALENTS

WOOD		EQUIVALENT GALLONS OF FUEL OIL	EQUIVALENT CUBIC FEET OF NATURAL GAS
lbs./cu.ft.	Btu/cord		
20	12,286,400	88	12,286
25	15,358,000	110	15,358
30	18,429,600	132	18,430
35	21,501,200	154	21,501
40	24,572,800	176	24,573
45	27,644,400	197	27,644
55	33,787,600	241	33,788

HOW FUEL WOODS COMPARE. The following table lists 109 woods considered commercially important by the U.S. Forest Service. Weights are based on 1 cubic foot of each wood at 12 percent moisture content. Calculations for the high-heat value per cord are explained in the foregoing text. Moisture contents for heartwood and sapwood are those you can expect in green, freshly cut wood. *Note:* In practice, the best wood-burning stoves and furnaces achieve only 50 to 70 percent efficiencies from the high-heat Btu values shown. So adjust your expectations accordingly. Highly resinous softwoods may produce as much as 6 percent more Btu than shown here, but resin contents vary widely within species. So the figures shown are reliable high-heat values for softwoods, followed by a plus sign (+), indicating that resins may increase Btu value anywhere from 1 to 6 percent.

HARDWOODS

TYPE OF TREE	POUNDS/CUBIC FOOT AT 12 PERCENT MOISTURE	BTU/CORD AT 12 PERCENT MOISTURE	AVERAGE PERCENT MOISTURE WHEN GREEN	
			Heartwood	Sapwood
Alder, red	25½	15,665,160	—	97
Ash:				
Black	30½	18,736,760	95	—
Blue	36	22,115,520	—	—
Green	35	21,501,200	—	58
Oregon	34½	21,194,040	—	—
White	37½	23,037,000	46	44

TYPE OF TREE	POUNDS/CUBIC FOOT AT 12 PERCENT MOISTURE	BTU/CORD AT 12 PERCENT MOISTURE	AVERAGE PERCENT MOISTURE WHEN GREEN	
			Heartwood	*Sapwood*
Aspen:				
Bigtooth	24	14,743,680	95	113
Quaking	23½	14,436,520	95	113
Basswood, American	23	14,129,360	81	133
Beech, American	40	24,572,800	55	72
Birch:				
Paper	34½	21,194,040	89	72
Sweet	40½	24,879,960	75	70
Yellow	38½	23,651,320	74	72
Butternut	23½	14,436,520	—	—
Cherry, black	31	19,043,920	58	—
Cottonwood:				
Balsam poplar	21	12,900,720	—	—
Black	22	13,515,040	162	146
Eastern	25	15,358,000	—	—
Elm:				
American	31	19,043,920	95	92
Rock	39	23,958,480	44	57
Slippery	33	20,272,560	—	—
Hackberry	33	20,272,560	61	65
Hickory, pecan:				
Bitternut	41	25,187,120	80	54
Nutmeg	37½	23,037,000	—	—
Pecan	41	25,187,120	—	—
Water	38½	23,651,320	97	62
Hickory, true:				
Mockernut	45	27,644,400	70	52
Pignut	46½	28,565,880	71	49
Shagbark	45	27,644,400	—	—
Shellbark	43	26,415,760	—	—
Locust, black	43	26,415,760	—	—
Locust, honey	37½	23,037,000	—	—
Magnolia:				
Cucumbertree	30	18,429,600	80	104
Southern	31	19,043,920	80	104
Maple:				
Bigleaf	30	18,429,600	—	—
Black	35½	21,869,792	—	—
Red	33½	20,579,720	—	—
Silver	29	17,815,280	58	97
Sugar	39	23,958,480	65	72
Oak, red:				
Black	38	23,344,160	76	75
Cherrybark	42½	26,108,600	—	—
Laurel	39	23,958,480	—	—
Northern red	39	23,958,480	80	69
Pin	39	23,958,480	—	—
Scarlet	41½	24,494,280	—	—
Southern red	37	22,729,840	83	75
Water	39	23,958,480	81	81
Willow	43	26,415,760	82	74
Oak, white:				
Bur	40	24,572,800	—	—
Chestnut	41	25,187,120	—	—
Live	55	33,787,600	—	—
Overcup	39	23,958,480	—	—
Post	41½	24,494,280	—	—
Swamp chestnut	41½	24,494,280	—	—
Swampy white	45	27,694,400	—	—
White	42½	26,108,600	64	78

TYPE OF TREE	POUNDS/CUBIC FOOT AT 12 PERCENT MOISTURE	BTU/CORD AT 12 PERCENT MOISTURE	AVERAGE PERCENT MOISTURE WHEN GREEN	
			Heartwood	*Sapwood*
Sassafras	28½	17,508,120	—	—
Sweetgum	32½	19,965,400	79	137
Sycamore, American	30½	18,736,760	114	130
Tanoak	36	22,115,520	—	—
Tupelo:				
Black	31	19,043,920	87	115
Water	31	19,043,920	150	116
Walnut, black	34½	21,194,040	90	73
Willow, black	24	14,743,680	—	—
Yellow poplar	26	15,972,320	83	106

SOFTWOODS

TYPE OF TREE	POUNDS/CUBIC FOOT AT 12 PERCENT MOISTURE	BTU/CORD AT 12 PERCENT MOISTURE	AVERAGE PERCENT MOISTURE WHEN GREEN	
			Heartwood	*Sapwood*
Bald cypress	28½	17,508,120+	121	171
Cedar:				
Alaska	27½	16,893,800+	32	166
Atlantic white	20	12,286,400+	—	—
Eastern red	29	17,815,280+	33	—
Incense	23	14,129,360+	40	213
Northern white	19	11,672,080+	—	—
Port Orford	27	16,586,640+	50	98
Western red	20	12,286,400+	58	249
Douglas fir:				
Coast	30	18,429,600+	37	115
Interior West	31	19,043,920+	—	—
Interior North	30	18,429,600+	—	—
Interior South	28½	17,508,120+	—	—
Fir:				
Balsam	22½	13,822,200+	—	—
California red	23½	14,436,520+	—	—
Grand	23	14,129,360+	91	136
Noble	24	14,743,680+	34	115
Pacific silver	27	16,586,640+	55	164
Subalpine	20	12,286,400+	—	—
White	24	14,743,680+	98	160
Hemlock:				
Eastern	25	15,358,000+	97	119
Mountain	28	17,200,960+	—	—
Western	28	17,200,960+	85	170
Larch, western	32½	19,965,400+	54	110
Pine:				
Eastern white	22	13,515,040+	—	—
Jack	27	16,586,640+	—	—
Loblolly	32	19,658,240+	33	110
Lodgepole	25½	15,665,160+	41	120
Longleaf	37	22,729,840+	31	106
Pitch	32½	19,965,400+	—	—
Pond	35	21,501,200+	—	—
Ponderosa	25	15,358,000+	40	148
Red	28½	17,508,120+	32	134
Sand	30	18,429,600+	—	—
Shortleaf	32	19,658,240+	32	122

TYPE OF TREE	POUNDS/CUBIC FOOT AT 12 PERCENT MOISTURE	BTU/CORD AT 12 PERCENT MOISTURE	AVERAGE PERCENT MOISTURE WHEN GREEN	
			Heartwood	Sapwood
Slash	37	22,729,840+	—	—
Spruce	27½	16,893,800+	—	—
Sugar	22½	13,822,200+	98	219
Virginia	30	18,429,600+	—	—
Western white	23½	14,436,520+	62	148
Redwood:				
Old-growth	25	15,358,000+	86	210
Young-growth	22	13,515,040+	—	—
Spruce:				
Black	25	15,358,000+	—	—
Engelmann	22	13,515,040+	51	173
Red	25½	15,665,160+	34	128
Sitka	25	15,358,000+	41	142
White	25	15,358,000+	—	—
Tamarack	33	20,272,560+	49	—

Costs for wood. Costs vary widely throughout North America. Prices for the best woods in wilderness regions may be lower than for the poorest woods in metropolitan areas. Six-inch logs of 1-foot length may cost $1 each in a large city. At that price per stick, you'd be paying nearly $600 per cord. On the other hand, a mixed cord of black ash and paper birch may cost only $15 in northern Minnesota, if you load your own in the supplier's yard. Where I live, a cord of mixed hardwoods costs between $70 and $100. In New York City suburbs, the same wood costs more; in the hills of Vermont, less.

Basically, costs depend on availability and popularity of a wood type. If you buy in spring, prices may be far lower than when the heating season approaches. You can usually negotiate discounts if you make the pickups yourself. Here, the more handling and transportation costs for the supplier, the more you'll be socked.

Beware of the dealer who delivers a cord of wood in a pickup truck. Most pickups are rated for ½-ton and ¾-ton capacities. That's 1000 to 1500 pounds. A cord of the lightest and poorest-burning woods weighs over 1800 pounds when well seasoned. Green, a cord of these poor woods would stress a pickup's suspension beyond its limits. A cord of medium-weight woods could destroy the truck. A half-cord of the heaviest woods would probably do the trick. Again, measuring a stacked cord is the surest way to get the volume you are paying for.

HOW TO DISTINGUISH TYPES OF STACKED WOODS. Usually, cords of heavier, better-burning woods command higher prices than cords of lesser woods. Seasoned logs, cracked and dry-looking at the ends, will normally cost more than green woods of the same type. But it's often difficult to distinguish among types of ash or types of maple, for example, even though their heating values may vary by more than 30 percent. Then too, a lot of smooth-bark aspen, with a potential of 14½ million Btu per cord, has been sold as smooth-bark beech, which can yield over 24 million Btu. A dealer may tell you he's delivering sugar maple, with 24-million-Btu potential, when he's really unloading 18-million-Btu silver maple. Let's give the dealer credit for honesty; maybe he just doesn't know his woods.

Experts can be fooled when buying stacked woods. The surest way to identify a tree is to use a field guide that describes leaves, fruits, seeds, winter buds, bark, wood texture, and weight. But when you see logs stacked or have them delivered, you seldom see more than trunks and limbs. In this case, a tree guide may not be much help.

Bark on young trees of many species looks different from that of older trees, and bark often looks different near the base of the tree than it does higher up. Red maple and sugar maple look a lot alike, but red maple is a far inferior burning wood. Some of the species of ash may be hard to distinguish until you burn them. Most oaks are tough to distinguish, but if you have any oak you've got good burning wood.

Yet some trees do have distinctive bark. Shagbark hickory, the birches, the cedars, and the pines head the list. Elms may be easily distinguished from other trees by their reddish heartwood and the galleries made by beetles under the bark.

You may never need to know more than a dozen types of wood, anyway, because those dozen may be abundant and highly popular burning woods in your area. Yet there's something about wood burning that fires an interest in woods, far beyond their Btu potential. A wood burner gets to know woods by their weight and feel when hefted, by the ways they respond to a bow saw, by the ways they split and season.

If, as a result of burning wood, you find yourself increasingly attracted to trees, don't be alarmed. It's a common phenomenon. Many wood burners find themselves pleasantly beset with a new tree consciousness. Some carry the thing so far as to buy a tree identification guide, plant seeds and seedlings, and learn how to prune and doctor trees for beauty and long life.

REGIONAL GUIDE TO NORTH AMERICAN FUEL WOODS. It's good to know that the hickories weigh about 45 pounds per cubic foot air-dry, and that they are among the best fuel woods, yielding about 27 million Btu per cord. But if you live in Arizona, or Idaho, or Alaska, you'll search in vain after hickories. The same applies to many kinds of trees with limited ranges in North America. The Forest Service list of important trees provides useful references. But what you, as a wood burner, need to know is what woods are available in your area.

For this information, I wrote to state and provincial foresters as well as to university extension foresters in all states and provinces. The idea was to get the views of professional insiders, nearly all of whom turned out to be wood burners themselves. Indeed, people do heat with wood in all parts of North America, including all of the southernmost states and—would you believe?—Hawaii.

The regional groupings shown on the accompanying map roughly follow those developed by the U.S. Department of Agriculture for ornamental trees. The boundaries themselves may be challenged because ranges of many trees jump boundaries. But for practical purposes, the chief types of wood fuel in each of the regions tend to give the regions their own wood-burning identity.

Northeast (22 states). Trees of high availability and high to medium Btu value include green, white, and black ash; beech; paper, sweet, and yellow birch; the elms; nearly all fruitwoods, especially apple and black cherry; hawthorns; flowering dogwood; the hickories; black and honey locusts; sugar and black maple;

the white-oak group; and the red-oak group. Of low availability but tremendous Btu value is American hornbeam (ironwood).

Trees of high availability but only medium to low Btu value include aspen; basswood; cottonwood; American elm; silver, red, and boxelder maples; yellow poplar; hemlock; and the pines.

The toughest-splitting woods include elms, beech, sycamore, apple, and sweetgum. Softwoods tend to produce creosote.

Southeast (9 states). Trees of high availability and high to medium Btu value include green and white ash; beech; river birch; American and slippery elm; nearly all fruitwoods; hawthorns; flowering dogwood; the hickories; black and honey locusts; sugar maple; the white-oak group; the red-oak group; persimmon; American hornbeam (ironwood); and eastern hornbeam. The pecan is available along the Mississippi River. Black-walnut limbs are excellent (sawmills pay high prices for trunks).

Trees of high availability but only medium to low Btu value: sweetgum; silver, red, and boxelder maples; southern pines; and yellow poplar.

The toughest-splitting woods include beech, elms, sweetgum, and sycamore Pines tend to produce creosote.

The Plains (6 states). Trees of high availability and high to medium Btu value include green and white ash; fruitwoods; black and honey locusts; oaks; and American and slippery elms. In the southern portion of the range, osage orange, pecan, persimmon, and American hornbeam (ironwood) make good fuel, though they are not abundant. Mesquite is a favored wood in Texas.

Trees of high availability and only medium to low Btu value include aspen in the Dakotas; silver, red, and boxelder maples; basswood; cottonwood; sweetgum; sycamore; eastern red cedar; hackberry; and, in limited pockets, pines.

U.S. Wood Fuel Regions

The toughest-splitting woods include elms, sweetgum, sycamore, fruitwoods, and mesquite. Pines tend to produce creosote.

The Rockies (8 states). This region is unique in the lower forty-eight states because of its comparative lack of hardwoods with good Btu value. Only Arizona has an abundance of these hardwoods, including gambel and emory oaks, as well as mesquite. Of the many softwoods in the entire region, only a few basic groups weigh over 30 pounds per cubic foot and yield over 19 million Btu per cord. These include western larch, Douglas firs, and pinyon pines. These heavy softwoods offer roughly the same Btu as hardwoods such as elms and lightweight maples, which are far scarcer.

The pinyons are essentially in the south and the western larch in the north. The Douglas firs are throughout the Rockies. Also abundant and offering good heat values are the junipers and the ponderosa and lodgepole pines.

Among the most difficult woods to split are elm, grand fir, western hemlock, junipers, pinyon pines, and the oaks. Those tending to cause the most creosote are western larch, Douglas fir, and ponderosa pine.

California. Trees of high availability and Btu value include eucalyptus, live oak, black oak, and Pacific madrone. Trees offering medium Btu value include tanoak, California laurel, Douglas fir, and ponderosa pine. Of all, eucalyptus is the toughest-splitting.

Oregon and Washington. Highest in Btu and high in availability is the Oregon white oak. Other popular hardwoods include red alder and bigleaf maple. Prime softwoods used for fuel are Douglas fir, western hemlock, and ponderosa pine. The softwoods tend to produce the most creosote.

Hawaii. Believe it or not, woods are used for heating in the cooler parts of the islands. Kiawe is highly available and also produces good cooking coals. Less commonly available but offering high Btu values are mamani and ohia.

Alaska. Best hardwoods are paper (canoe) birth and alder. Best softwoods include American larch (tamarack) and western hemlock. Highly available but with low to medium Btu are aspen, cottonwood, and white spruce.

Canada. Principal types of trees in Canada tend to follow transcontinental ranges, except near the U.S. border. The ranges are best shown in the accompanying map and legend. Basically, though, the populous areas of southeastern Canada hold most woods noted earlier for the northeastern U.S.

PRINCIPAL TREES OF CANADA

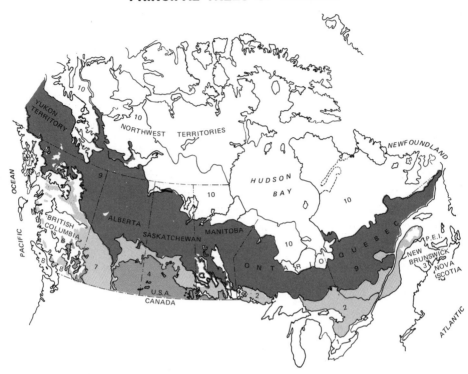

Range   Principal Trees

1.   Beech, maples, black walnut, hickory, oaks

2.   Oaks, maples, yellow birch, paper birch, red and white pines, eastern hemlock

3.   Maples, yellow birch, paper birch, red oaks, some white oaks, red spruce, balsam fir

4.   Aspen, willow, bur oak, paper birch, jack pine, white spruce, American larch (tamarack)

5.   Aspen, willow

6.   Aspen, Douglas fir, lodgepole and ponderosa pines

7.   Engelmann spruce, alpine fir, lodgepole pine, western red cedar, western hemlock, Douglas fir, paper birch

8.   Maples, western red cedar, western hemlock, Sitka spruce, Douglas fir, some white birch

9.   Aspen, white birch, white spruce, black spruce, balsam fir, jack pine

10.   White spruce, black spruce, tamarack, some white birch

13 | Seasoning Wood

THE LOWER A WOOD's moisture content, the better the wood will burn. As shown in the previous chapter, free water in freshly cut "green" wood may weigh anywhere from 40 percent to more than 150 percent of the wood's oven-dry weight. A moisture content of 100 percent indicates that the wood is half water by weight.

BURNING GREEN WOODS. It's best to avoid burning woods green. Woods with a moisture content of over 50 percent just don't burn well. They are difficult to ignite. Their free water consumes Btu as the water is boiled off, and the water may consume so much heat in vaporizing that the temperatures in the wood's inner layers may not be sufficient to produce combustion gases. Thus very wet wood may not be capable of supporting its own flames. In this case the heat from drier woods in the fire chamber is needed until the water in the green wood boils off.

Also, wet wood's lower Btu value results in less heat transfer to the house. If you burn green woods of high moisture content, you'll notice that the wood doesn't really provide a lot of room heat until it's done sniveling and hissing. Besides, the water vapor caused by green wood has a cooling effect in the flue. Relatively cool flues of 150°F to 500°F tend to condense water vapors and volatiles that make up creosote.

So if you must burn green woods, try to pick those with low moisture contents on the stump. Some of these are noted in the previous chapter. Then, since heartwood is generally drier than sapwood, you can split out the heartwood and burn it first.

PREVENTING ROT. The drying of wood also prevents rotting that results from bacteria and fungus. After felling, the sapwood of most types of wood is more susceptible than heartwood. Heartwood of some woods is very resistant to rot; resistant woods include cedars, black cherry, junipers, black locust, mesquite, oaks in the white-oak group, osage orange, sassafras, and Pacific yew. Highly susceptible woods include alder, ashes, aspens, basswood, beech, birches, cottonwood, elms, hickories, maples, oaks in the red-oak group, yellow poplar, spruces, sweetgum, true firs, and willows.

As wood rots, its nutrients break down into compost, which is great for enriching the soil. Other by-products include carbon dioxide, water vapor, and heat—the same as those that result from burning the wood. But the loss of heat amounts to a loss of potential fuel value as the wood rots. So it's smart to prevent rot in your firewood. Fungi and many kinds of bacteria fare best in damp conditions, with temperatures ranging from 60°F to 90°F. It follows that summer temperatures hasten rotting. Also, when wood is in contact with the ground, it absorbs ground moisture. Repeated soakings from rain also help maintain damp, rot-favoring conditions.

As it turns out, the techniques for fighting rot are the same as those for lowering a wood's moisture content—a process called seasoning.

SEASONING: TWO OBJECTIVES. First, you need to promote the removal of existing moisture. Second, you need to prevent the wood's gaining moisture from rain and snow, as well as from the ground.

Removing moisture. Here you're concerned with four major factors: (1) the amount of air circulating in the wood pile, (2) the size of the wood pieces, (3) the amount of bark on each piece, and (4) the air temperature. Let's look at each of these factors.

Air circulation. The more air circulating around each piece of wood, the better. Each waft of air tends to draw moisture from the wood until the wood's moisture content is in equilibrium with the air's. From then on, the wood will absorb and lose moisture in response to changes in air humidity. Thin pieces of wood are more quickly affected than thick pieces.

It's important to note that relative humidity in air is expressed as a percent of what the air can hold at a given temperature. This relative humidity is always far higher than the percent of water in air-dry wood. For example, the average relative humidity in the U.S. Southwest ranges between 20 and 40 percent. Air-dry wood in equilibrium with these humidity levels will have moisture contents ranging from 4 to 8 percent. In the Pacific Northwest and along the East Coast, average relative humidity is about 80 percent, and this will give air-dry wood an average moisture content of about 16 percent. So in dry climates, it's possible to season wood nearly bone-dry. In humid climates, you may be able to achieve 15 percent moisture levels, at best, during only the driest months.

In the Northeast, veteran wood burners suggest that wood should season six months under cover for acceptable burns, one year for burns near potential, and two years for best burns.

To provide the best circulation around each stick of wood, you should stack the wood log-cabin style—creating a chimney with a hollow center and plenty of space between layers. The stacks themselves should be separated enough so they don't screen one another from drying breezes.

Size of wood pieces. The smaller you split wood sticks, the closer the free moisture is to the surface. Circulating air then laps up available moisture migrating from the center of the wood, and the now drier wood on the surface absorbs more moisture from the inner layers. Air circulates past the surface again, and absorbs more moisture. This continues until the wood moisture level reaches equilibrium with the air's relative humidity.

So if you want to dry wood rapidly, split it into thin pieces. However, thin pieces of air-dry wood tend to burn faster than thicker pieces, because the heat of the flames is concentrated in a small volume of wood. Inner layers of thick wood heat up slower, and so release combustible gases slower too.

Wood bark. Bark protects the delicate tissues of live trees and helps retain moisture, and it continues these functions after the tree is felled. Especially tight and resinous barks, such as those of the birches, retain moisture so well that

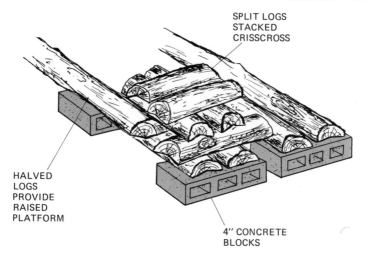

SPLIT LOGS
STACKED
CRISSCROSS

HALVED
LOGS
PROVIDE
RAISED
PLATFORM

4" CONCRETE
BLOCKS

Log-cabin stacking of split wood provides the best circulation. Concrete blocks and halved logs provide a raised platform that promotes circulation and prevents ground moisture from wicking up and inducing rot. This year's platform logs can be saved and burned near the end of the heating season or else burned the next year.

birch may rot badly in a single year if left unsplit. Other barks are more porous and may even crumble away after a short time. But all barks help the wood retain moisture to some degree.

For rapid seasoning it's wise to split all logs over 3 inches in diameter. Larger logs should be split to burning size immediately rather than merely being halved for later splitting.

Then if the wood has no artificial cover, you should stack it bark side up, if possible, so that the bark can shed rain.

Air temperature. Summer months promote fastest drying. Warm ambient air has greater capacities to absorb moisture, and sun-heated wood promotes that absorption. Yet even on cold winter days, wood can continue drying. In this case snow and ice on the wood may sublimate—that is, they may pass directly from solid to vapor, with no intermediate melting stage.

Preventing moisture gains. The importance of a cover depends on where you live. In the hot arid Southwest, a cover may retard drying. In this case, a cover may shed only occasional rains, but on sunny days it may shade the wood from drying effects of the sun. But in areas with moderate to heavy precipitation, a cover pays dividends—especially during the last few months before burning. Many people leave their wood stacked in the woodlots where the trees are felled. The wood may lose one-half to one-fourth of its weight in this seasoning. Then it's lighter to handle and transport to shelter nearer the home.

Many of the better-burning woods rot rapidly when they are in contact with the ground, so ground moisture can be a greater problem than soaking from precipitation. At the very least, the wood should be elevated on tiers of less desirable burning wood. Concrete blocks are better, though, because they don't wick ground moisture up to the stacked wood, as wood tiers may. Elevating the wood also minimizes the amount of rain splash on bottom logs as well as the effects of drifting snow, and by raising the wood off the ground, you promote ventilation and discourage creatures such as rodents, ants, and termites.

Stacking, covers, and sites. There are two divergent schools of thought on woodpiles. In one, a proper woodpile is any stack, heap, or scattering of logs and sticks wherever there's space enough. In this case, the wood usually lies right where it was split. Time is money to many people, and stacking means more handling time. So, no stacking.

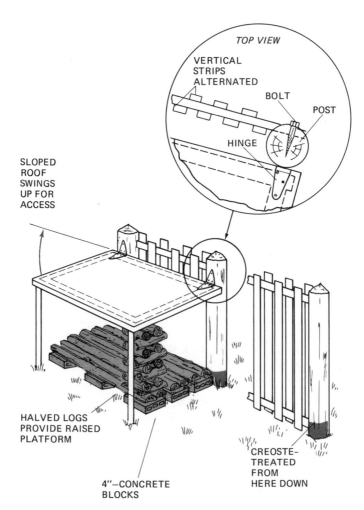

This woodpile-fence serves two purposes. It screens off the work area, which can get cluttered at times, and it provides support for a solid lean-to roof. Vertical slats of the fence are mounted alternately on the horizontal members to provide air circulation as well as the screening. The hinged roof allows easy access, and it may be raised to admit the sun's rays.

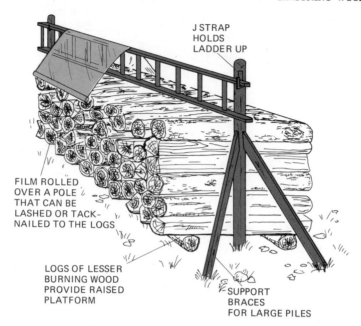

J STRAP
HOLDS
LADDER UP

FILM ROLLED
OVER A POLE
THAT CAN BE
LASHED OR TACK-
NAILED TO THE LOGS

LOGS OF LESSER
BURNING WOOD
PROVIDE RAISED
PLATFORM

SUPPORT
BRACES
FOR LARGE PILES

If you cover wood with polyethylene film, find a means of elevating the film so that air can circulate freely. Otherwise moisture evaporating from the wood will condense on the underside of the film and drip onto the wood. The ladder gimmick does the trick, while giving you a place to store your ladder. Then if you need the ladder for something else, you can easily peel back the film for the day. Variations on this theme are endless.

Cross stacking can provide buttresses for unusually tall tiers of logs, as shown. Advantages of this method of stacking are excellent air circulation and ease of placing and retrieving the logs. This pile belongs to Frank Smith of Sherman, Connecticut.

This wood shelter was twenty years old at the time of the photo. It has survived winters with over 3 feet of snow and temperatures under 40°F below. It covers about two full cords of wood that are first seasoned in uncovered mounds, shown in the background. The roof is of corrugated tin and roofing paper.

In the second school, a proper woodpile is as much an indicator of the taste and preparedness of the residents as their outbuildings and general landscaping. Here, the woodpile becomes a vital element in a scene that makes people stop to take photographs. Woodpile artisans have fashioned geometric extravaganzas: octagonal piles, pentagonal piles, piles with flying buttresses, piles that resemble Quonset huts, piles that form a maze, and piles that become fences along property lines.

Yet as woodpiles become more fanciful, they become harder to cover. Indeed, the cover tends to conceal the creation and so may be waived for the sake of art. Still, in most parts of North America, a cover should be applied at least before the descent of freezing rains and snow. For if every log is cloaked in a half-inch mantle of ice, the ice must be chipped off or melted off before the wood is stuffed into the fire. And many is the woodpile artisan who spurned a cover and wound up attacking his ice-locked pile with a sledge and bar.

Many veteran wood burners prefer conventional shed roofs for cover. These range from peaked woodsheds with vented walls all the way to lean-to roof extensions from existing outbuildings. Then too, lean-to roofs can be extended from fenceposts.

Many people simply drape 4- to 6-mil polyethylene plastic over the pile. With $5 to $10 worth, you can cover three to six cords. But this plastic is short-lived. Sunlight makes it so brittle that it may crack and rip to tatters in a year. Even when the plastic is new and supple, sharp wood edges can puncture it if you're not careful when draping it. This plastic has another drawback: It's not biodegradable or recyclable. It becomes just one more petroleum-based product for landfill. *Note:* Always allow space for air circulation between wood and cover.

Nylon and canvas tarps last longer than polyethylene film. But nylon is another nonrecyclable. Both fabrics will cost you $15 to $25 for enough to cover a single cord.

In the long run more permanent shelters have advantages. There may be tax advantages in extending lean-to type roofs from existing buildings rather than erecting new buildings. But by all means, avoid extending the shelter from your house. Woodpiles next to houses may lead to insect invasions.

WOOD AND INSECTS. I once downed an old yellow birch in my backyard. As the 42-inch bow saw reduced the trunk to splittable lengths, I discovered a large irregular hollow inside—aswarm with carpenter ants. They were miffed, but there was no way I could make amends. I was still a fledgling woodcutter then and bore no grudges against any creature, so I felt guilty. But there were plenty of trees in the nearby watershed, so I reasoned the ants would have no trouble finding a new home.

They found a new home all right—between the insulation and rafters over our son's bed. We didn't suspect a new nest even though we'd been swatting occasional foragers in our kitchen for some months. But late one night, we heard soft scraping sounds overhead. I removed one ceiling tile, noted a couple of ant sentries, and replaced the tile.

When the exterminator arrived, he considerately tried to calm us by explaining that sightings of ants were not proof of a nest in a house. Ants usually forage at night for fruits, other sweets, and grease before returning to a nesting tree or within woodwork. He removed the loose ceiling tile, and three sentries emerged to gather news. The exterminator mounted a chair and probed with a flashlight. He dismounted and said gravely, "There's conclusive evidence of a nest."

He explained that he'd seen many more ants and some small amounts of sawdust. Like termites, carpenter ants prefer decayed or damp wood. Indeed, this nest was centered right where we'd patched for several roof leaks.

The ants don't eat the wood, as termites do, but excavate galleries for the queen's eggs. These galleries can weaken the wood until it collapses. The galleries are made across the wood grain, as though done with a fitful round-nosed gouge. The wood scrapings range from fine to coarse and simply fall from the excavations. Damp wood provides ants with moisture their bodies need, but it also makes for easier excavating than dry wood. Ants will expand to dry wood after exploiting the damp wood, but they need damp wood to thrive. So they tend to nest where water leaks have occurred, such as under roofs, sinks, and toilets. Or they may nest in any wood that is damp from humidity and poor ventilation.

Carpenter ants play a useful role in nature and lead an interesting social life. But they can destroy house framing, as this photo evidences. Note that these galleries were excavated in moldy, damp wood that begged for an ant infestation.

You can start an ant infestation by cutting down a nesting tree, as I did, forcing a massive migration. Or you can carry a queen indoors in your firewood.

Our exterminator estimated that carpenter ants generate twice as much of his business as termites do. But we live in the North. In warmer coastal regions of North America, termites account for an increased share of the extermination trade.

Termites. There are two basic types: subterranean termites and dry-wood termites. Subterranean termites live underground and prefer damp, warm soils where there's access to damp wood. They don't nest in the house. Instead, they commute. Trapped indoors away from the ground or other source of moisture, they die. Earthen shelter tubes emerging from the ground and routed toward wood timbers are sure signs of infestation in a house. Wood damage may not be apparent until you insert a knife or tap the wood with a hammer. Unlike ants and dry-wood termites, subterranean termites form channels that follow the wood grain. The sides of these channels are covered with small grayish-brown specks of excrement.

The other termites—dry-wood termites—don't need contact with the ground. They are as content with dry or wet firewood as with wood indoors. They may nest in a home after being carried indoors in the firewood. In east-central, southern, and west-central coastal regions where dry-wood termites are most common, it pays to examine firewood carefully before bringing it indoors. The U.S. Department of Agriculture advises you to look for "clean cavities cut across the grain in comparatively solid, dry wood. These cavities contain slightly compressed pellets of partially digested wood. Some of the pellets are pushed through tiny openings in the exterior where they often form in piles on the surface below. This termite also seals its entrance holes with a brownish-black, paper-thin secretion which may contain a few of the pellets."

Elm Bark Beetles. These carriers of fungal spores that kill elms pose no threat whatever to house structures. The beetles feed in the crotches of elm branches that are two to four years old. They breed under the bark of damaged, cut, and dead elms. If you have elm wood in your woodpile, strip off the bark immediately. This will stop the incubation of any eggs and the growth of larvae, and it will prevent future breeding under the bark. Then burn the bark. If some beetles should enter the house with your firewood, they'll be mere nuisances, deserving a swat.

Other insects. Roaches, beetles, centipedes, and spiders may also enter with fuel wood. Most are mere nuisances. But roaches can become a health hazard as well.

Preventing insect invasions. Many people ring the base of ant trees with poison. But since insecticides often enter food chains, poisons should be only of the short-lived type. *Organic Gardening* magazine claims successes with bands of cotton "made sticky with Tanglefoot or Stikem." Wrapped around the trunk, the sticky cotton could catch all foragers. But to get the queen and the nest itself, it's probably best to down the tree and saw out the trunk section containing the nest before dousing the nest with kerosene and burning it.

Since ants and subterranean termites tend to abandon dry wood, let your firewood become air-dry before you burn it. Even then, you should avoid storing more than a half-day's worth or so indoors. In cold weather, wood can be stacked near the house for easy access. But during warm weather, when insects become active, the wood should be stored more than 25 feet from the house, if possible. Again, the wood should be stacked on non-wood tiers, because wood tiers attract both ants and subterranean termites.

Other preventive measures. Remove all rotting wood and stumps near your home. Repair roof and plumbing leaks promptly. Ensure adequate ventilation in attics and basements.

It might also pay to invite an exterminator to inspect your home. Often there's no charge for the visit. At that time the exterminator can show you where to watch for signs of invasion as you burn firewood over the years, and he may even suggest where you can patch or caulk potential points of entry along your foundation and walls.

14 | Hand-saw Basics

A FELLOW ONCE regaled me with tales of his wonderful new chain saw. It seems he had used the machine to prune his apple tree. What convenience! What efficiency! As conversation drifted to what else was new, the somewhat portly fellow offered that he'd been attending exercise classes at the local gym. When I inquired if there wasn't an inconsistency in attending exercise classes and pruning with a chain saw, the fellow emitted a few snorts and changed the subject.

We have ovens that don't make us bend. We have TV channel selectors that let us remain on our duffs. We have lawnmowers that power themselves—even transport us forth and back. We have chain saws for pruning little trees and cutting less than a cord of wood a year. And we talk of workouts at the gym. Is this an inevitable perversion?

Hand saws can provide a body- and mind-stimulating regimen. Hand saws are inexpensive compared to chain saws. They require little to no maintenance. They are quiet to use and cause no air pollution. Depending on the amount and types of wood you cut, a hand saw may or may not require more time than a chain saw. But it will always require more effort.

Good chain saws can fell trees fast and saw up a cord of any size wood in an hour or two. Using a two-man crosscut saw, a couple of champion lumberjacks can outcut a standard chain saw in sprints of a minute or two. But when the sweat begins raining from their chins, the lumberjacks must pause for a breather. Then the chain saw catches up and pulls ahead. Whether electric- or gas-powered, chain saws are noisier than hand saws. Noise from gas engines can damage unprotected ears, and noise from chain teeth on wood from either type of chain saw can reach an ear-damaging 90 to 100 decibels. The electric saw may pollute air near a distant power plant, while the gas saw pollutes air where you use it. For every day you use a power saw, you should figure on devoting an extra hour to maintenance and sharpening. And power saws are far more dangerous—both from standpoints of mechanical failure and user carelessness.

TYPES OF HAND SAWS. There are several types of hand saws, including some excellent general-purpose types as well as special-purpose types.

Two-man crosscut. This is the saw that felled most of the virgin timber in North America. With upright wooden handles bolted at each end, the two-man is available today in lengths from 5 to 6 feet. Costs range from $45 to $60 new, but you may be able to pick up a fine old relic for a few bucks at a garage sale. Lumberjack festivals often include a contest that sees only one man sawing with this two-man creation. But the act requires superstar endurance and an ability to keep the far end of the saw from bull-whipping. If you're the type who plays squash or racketball regularly, or if you jog five miles and lift weights as a regimen, you might enjoy putting up firewood with a two-man crosscut and a partner of your equal. This saw should be sharpened daily, though, unless your lust for exercise is unbounded.

One-man crosscut. This looks like an extra-long carpenter's saw and may have a hole bored in the narrow end so that you can attach a handle for a helper. Lengths range from 3 to 4 feet. Costs run from $20 to $35. This saw is ideal if you like exercise for its own sake but don't have a partner with your enthusiasm. The blade is about as thick as a penny, and the teeth are set to cut a kerf a little wider. If you attempt to cut spring wood with the sap flowing, the wet wood tends to swell and pinch the 4-to-6-inch deep blade unless the teeth are set to cut a wide kerf. If one of these saws is left in its own kerf in fast-swelling spring wood for a few minutes, even King Arthur would have trouble extracting it.

Bucksaw. This is the thin-bladed saw tensioned by means of an H-shaped wooden frame drawn together at the top by a turnbuckle. Pioneers used twisted rawhide instead of the turnbuckle. In the old days, simple V-cut teeth needed frequent sharpening, but a sharp bucksaw could turn out plenty of stove-length

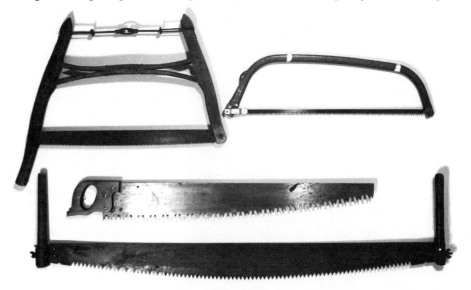

From bottom to top: The old-time two-man crosscut has old-style V-cut teeth, and its handles are secured by wingnuts. The one-man crosscut has the more modern cutter-and-raker teeth; note the holes near each end for mounting top handles. The old V-cut blade of the H-framed buck saw is tensioned by the turnbuckle at the top. The bow saw has a handle that is also the tensioning lever for the blade. Blades are available in various lengths and with cutter-and-raker teeth similar to those shown on the one-man crosscut saw or with groupings of V-cut teeth as shown on the bow saw. Hard-point Scandinavian blades can't be sharpened with conventional steel files because the file steel is softer. So most are simply discarded after a long, productive life.

wood. Today you can fit a bucksaw frame with a thin, long-lasting Scandinavian blade that may kindle your interest in pioneer crafts. If the bucksaw has a shortcoming, it's blade length. For in order to put enough tension on a longer blade, you need a taller and heavier H-frame. After a point a heavier frame with a higher center of gravity becomes unwieldy for one-hand operation. Still, a homemade bucksaw with blade up to 30 inches can be a handsome and efficient maker of firewood.

Bow saw. These are the most practical, efficient, and economical hand saws available. Many people call them Swedish bow saws. The rig employs a thin, replaceable blade tensioned by a lever at one end of a tubular steel bow. Cost for a quality 30-inch saw with blade is only about $10. Replacement blades cost only $3 to $5. I've found most blades of Scandinavian manufacture far superior to others. A good blade should last through 4 to 8 cords easily. A 42-inch saw will cost about $15. Shorter bow saws are available, but they are meant primarily for camping and pruning.

For bucking wood to stove lengths, a 30-inch bow saw is ideal on wood up to 4 inches in diameter. But for larger logs, a 42-inch saw lets you rear back with a two-hand stroke and a pendulum motion of your upper body. The bow saw has one limitation: its clearance between blade and bow. For example, if

For top bow-saw efficiency on logs up to 4 inches in diameter, lean directly over the work, bearing down mainly on the downstroke as demonstrated by Tom Grundvig. The saw-buck shown has cross members spaced at stovewood length; this helps in estimating desired log lengths for one-hand bucking, and it allows three cuts on a larger log without need to reposition the log. Cross members are held by bolts, allowing the buck to be closed for storage.

Tom Grundvig shows appropriate hand and body positions for bucking with a 42-inch bow saw. Note the light chain holding the log stable. Two bands of white tape on the bow aid in marking wood for desired lengths.

you are attempting to fell a tree with a 2-foot diameter, the bow of a 42-inch bow saw will be bumping the tree after the blade has penetrated 10 to 12 inches. This may not be enough clearance for you to fell a large tree or cut through the trunk of a fallen tree unless you notch the kerf with an ax to allow the bow to enter farther. Usually the bow is no obstacle when you are bucking on a sawbuck as long as you can turn the log for a second cut.

BUCKING WOOD ON A SAWBUCK. Most sawbucks employ two or more X-shaped pairs of poles or 2×4s that are held upright by horizontal members. Then the logs to be bucked are laid in the V formed by the tops of the Xs. The height of the cross in the Xs determines the ultimate height at which you must work. Best heights may vary, depending on the types of saws you use and the average size of logs. If you use both chain saw and bow saw you'll probably find that a cross just above your knees is about right. But it's wise to make the legs longer than you think necessary at first and then shorten them, if you like, after you've attempted some bucking.

One-handed bucking of smaller logs and branches works best when you can

Widths of saw kerfs are evidence of the comparative workloads placed on different saws. At far left is a ⅜-inch kerf made by a chain saw. Next is an ⅛-incher made by a one-man crosscut. Then comes a ¹⁄₁₆-incher made by a bow saw. Of the two hand saws, the bow saw needs to break through only about half as much wood fiber, and the sides of its shallower blade rub far less wood and cause less friction.

Homemade blade guards protect the teeth from accidental wear caused by rocks of metal, and—more important—prevent personal injury. These were made from ¼-inch plywood and spacer blocks. They can be held in place with twine or stretch cables.

stabilize the work with the free hand while sawing with the other. This puts your head almost directly above the work. The effort for this type of bucking is concentrated almost completely in one shoulder and arm, rather than throughout the body. So when one arm gets tired you can switch to the other. Here it's best to bear down a bit on the downstroke and let the weight of the saw do the work on the upstroke, letting the blade float back up. The technique takes about five seconds to master, and you'll know you've got it as soon as things feel right.

Two-handed bow-saw bucking. The bucking height you like for one-handed bucking will probably feel about right for two-handed bucking as well. Here you'll be whaling with both hands, arms, and shoulders, as well as your back, hips, thighs, and toes. Obviously, two-handed bucking tends to get more of your juices flowing than one-handed bucking does. It also quickly produces rapid heartbeat and a lot of panting. My guess is that two-handed bucking consumes nearly as much energy as jogging does. Slow, easy jogging and slow, easy bucking tire me about equally. Fast, hard bucking seems to me as strenuous as fast running. Yet you can stop as often as you like. There's no need to buck too fast or too long. There is a Zen to bucking with a bow saw that is revealed only to those who take it easily, rhythmically, breathing deeply, and getting into the experience. Pushups, chinups, and jogging are good conditioners.

Oh, yes, how about holding the log still? A heavy log will often be stable enough by virtue of its own weight. Unlike the one-handed method that includes bearing down on the downstroke, two-hand bucking should rely almost entirely on the weight of the saw as you float it in long even strokes through the kerf. Even so, medium-size logs want to jump around a bit, and this knocks you out of rhythm. To stabilize a log, just secure a light chain on the far side of the buck, run it over the log, and fasten it on your side. If that's not snug enough to do the trick, yank the top loop of the chain a few inches along the bark to pull it taut. On a smooth-barked log, you may have to drive a spike through one of the links into the log.

15 | Chain-saw Basics

HAND SAWS CAN put up a year's supply of firewood, all right. But chain saws make the work much easier. Chain saws are definite time and energy savers when you are bucking logs over 4 inches in diameter. They're super on really big stuff. Used on felled wood, they allow you to cut about three-fourths of the way through logs on the ground before rolling them and completing the cut—no lifting required, no lifting injuries.

But if you are just bucking a pile of logs in your backyard, a good chain saw can be slower than a bow saw or bucksaw when the following three conditions coincide:

• When the logs are less than 4 inches in diameter. Here small logs allow one-handed bucking, which causes little fatigue and allows you to switch arms if one becomes tired.

• When working time is limited to less than an hour. This may occur if you decide to devote only an hour each day to your woodpile. Hand bucking requires negligible setup and knockdown time, but chain-saw preliminaries and tidying up can easily waste fifteen minutes themselves. So if you plan to cut only for a short while, a hand saw can put you well into the work, long before you might fire up a chain saw.

• When you are working alone. By placing your sawbuck close to the woodpile, you can continue holding a bow saw with one hand while reaching for the placing of the next pole. This is unsafe with a chain saw.

TYPES OF CHAIN SAWS. Chain saws may be gas-powered or electric. Both types employ endless chains with cutter teeth, all driven around a grooved metal bar.

Gas chain saws have single-cylinder, two-cycle engines with piston displacements ranging anywhere from 1.6 to 8.36 cubic inches. The smallest engines are fitted with bars of 8 to 12 inches. The largest engines may link with bars over 5 feet long. The most popular and practical homeowner models have 2- to 4-inch displacements and carry bars ranging from 12 to 20 inches. Of course, the longer the bar, the greater its potential contact with great masses of wood. So to bust through lots of wood in single cuts, the most powerful engines with the largest displacements are matched to the longer bars. Gas saws may weigh anywhere from 7 pounds to 24 pounds, without bar and chain.

Electric chain saws employ the same chains and bars as gas saws. Electric saws are rated in kilowatt outputs. A small 1.1 Kw motor may be fitted with an 8-to-10-inch bar, a 1.9 Kw motor may be fitted with a 12-to-16-inch bar, and motors of 2 to 4 Kw may carry bars from 16 to over 30 inches. The largest motors are

146

designed for 220 volts rather than the 110 volts that smaller and medium-size electrics run on. Weights without bars range from 7 to 30 pounds.

Electric saws. For bucking in the backyard, electric saws deserve serious consideration. Here you should discount the cheapo electrics that are built light and may even have flimsy plastic housings. These are toys that you might as well throw away when something goes wrong, because repair costs would approach those of a new saw. Better electrics are double-insulated with fiber-filled nylon or polycarbonate. A quality electric will cost over $100, yet may still cost less than some of the cheap gas saws.

Electric motors need a change of brushes now and then. The gearbox needs lubrication and the cutter chain and bar need a continuous supply of oil, much as the cutting assembly of a gas saw does. But electrics don't burden you with all the maintenance and fuel pouring that gas saws need, nor do they produce noxious fumes. Electric motors are quieter than gas engines, though the cutting teeth of either type of saw can generate 90 to 100 decibels when smashing through hard, dry wood. That's noisy enough to cause ear damage after a time, so it calls for well-fitting ear plugs or sound-deadening muff-type protectors.

Before you get an electric, though, you should check to be sure you'll be able to cut near an adequate power outlet that protects you from shock by means of a ground-fault interrupter (GFI). Electrical codes now require that all new outdoor outlets be protected by GFIs anyway. So if you don't have a GFI for safety,

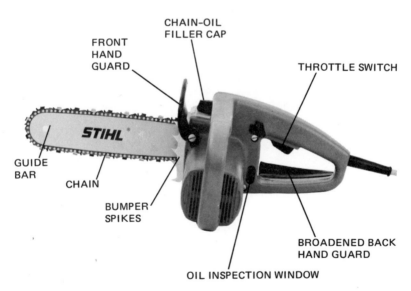

Stihl's E 10 electric has a fixed front hand guard, since there is no clutch that would allow a chain brake mechanism. The back hand guard is broad, providing protection against the possible whipping of a broken chain. Since electrics can be run indefinitely without running out of fuel, you don't have an empty fuel tank to remind you to refill the chain-oil tank, as is the case with gas saws. So you have to pay attention to the oil level shown in the inspection window.

you might as well have one installed. To ensure that you don't lose power, your extension cord should be a minimum of 14 gauge for short runs.

Old-fashioned electrics were plagued with motor burnouts when the chain bound up in wood. But most modern electrics have welded-wire rather than soldered-wire connections that won't melt when overheated. To my knowledge, only Skil makes an electric with a clutch that will disengage to protect the motor when the chain hits heavy going.

Safety note: Since most electrics do not have clutches, as gas saws do, the chain continues to rotate after you release the trigger, because of the inertia of the motor and gear train. Obviously, a moving chain after trigger release adds to the hazards.

Gas saws. For woods work and for all-around versatility, gas saws deserve the vote. Besides, gas saws are available from more makers than electrics are, and in

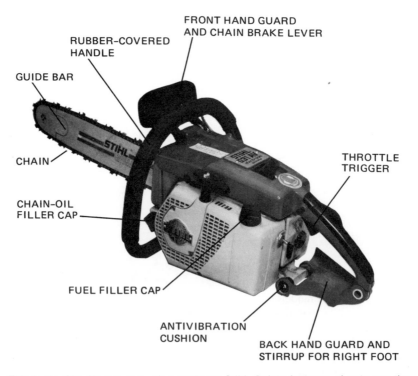

FRONT HAND GUARD
AND CHAIN BRAKE LEVER

RUBBER-COVERED
HANDLE

GUIDE BAR

CHAIN

CHAIN-OIL
FILLER CAP

THROTTLE
TRIGGER

FUEL FILLER CAP

ANTIVIBRATION
CUSHION

BACK HAND GUARD AND
STIRRUP FOR RIGHT FOOT

This is the 031 AV gas saw, also made by Stihl. Safety features of note are the combination front hand guard and chain brake lever, and the back hand guard. The guards protect the hands from the possible whipping of a broken chain as well as from sharp branch stubs. The front guard could prevent the front hand from slipping onto the moving chain. The rear hand guard also protects the saw controls from unwanted effects of branches. And, *very important*, the chain brake lever stops the chain if the front hand bumps the lever, as might occur during a kickback. Other features are noted later in this chapter.

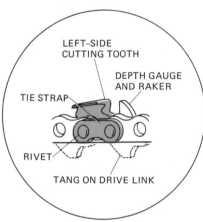

LEFT-SIDE
CUTTING TOOTH

DEPTH GAUGE
AND RAKER

TIE STRAP

RIVET

TANG ON DRIVE LINK

Gas and electric saws use basically the same chains and bars. Shown here are chain components fitted over a sprocket-nose bar by Poulan. Most makers offer bars without sprockets too—calling them hard-nose bars. Note the tang on the drive link; it carries oil from the saw reservoir along the groove in the bar. Some chains have added channels that force lubricant into rivet joints of the chain.

a far wider range of models and options. Also, the power units of some gas makes can be adapted to special attachments such as hedge trimmers, posthole diggers, and winches.

Gas saws burn only regular leaded gasoline mixed with special oil. You must never use gases with additives or oil with detergents and other additives because they turn to glue inside a two-cycle engine. Most American-made gas saws burn a gas/oil mix of 20 to 1. Most European makes burn a 40-to-1 mixture. But specs may vary among models by a single company, so be sure to follow the manufacturer's advice on mixtures. You should also add special oil to the chain-oil reservoir. When the saw will be idle for over a month, you should drain the fuel tank and then start the engine to burn all fuel out of the carburetor. Otherwise the stuff will turn to varnish. There's more to seasonal storage, but your owner's manual should cover the details.

RENTING A CHAIN SAW. Renting makes sense when you have several large trees to fell, or when you have several cords of wood ready for bucking. Rates may run anywhere from $20 to $40 a day, with a slightly higher charge for a whole weekend. The rate is roughly proportional to the saw's value and size.

Renting pays best returns if you can cram an entire year's worth of cutting into a day or two. This way a $25 day rental would give you roughly the equivalent of eight years' worth of use from a rental saw that would otherwise retail for $200. A $35 rental each year would give you roughly the equivalent of ten years of use from a $350 saw. Besides small fuel and oil costs and some touch-up sharpening, you'd be virtually free of maintenance tasks and service costs over the eight-to-ten-year period. Another bonus is that renting lets you test and compare saws for a few years before you select a model for purchase.

Yet drawbacks to renting can be significant. You may not be able to find a saw in top condition, or the saw may not have the right safety features or the right bar length and power for you, or the saw may take some getting used to before you can run it efficiently or even safely.

BUYING A SAW. The ownership ethic is a powerful one. As evidence, annual chain-saw sales have exceeded 2 million. All chain saws are dangerous tools, though some are less dangerous than others. All saws will cut wood, but some will cut better and last longer than others. So it pays to do some comparative shopping.

Used saws. Since new electric saws are available at comparatively low prices, and since electric motors may be on their last legs without your knowing about it, it's risky to buy a used electric unless you know it's had little use and good care. At the very least, ask for a demonstration, and then, if all looks okay, ask for a very low price.

Used gas saws are risks too, but if you are mechanically inclined, they may hold an attraction for you. Then again, you might find a bargain that will run as well as it did brand-new. Gas saws are subject to excessive wear resulting from improper fuels and oils. The cylinder and carburetor may have varnish buildups because the owner didn't drain the fuel tank and run the carburetor dry before storage. The chain and bar may not have been getting enough oil, or even the right kind of oil. If the owner will give you a demonstration, he may even let you take the saw to a chain-saw repair shop for a check. The shop's service fee will be worth it almost every time. If you buy, be sure you have the owner's manual.

New saws. Discount stores and direct-mail houses may offer saws at lowest prices, but there's a catch. When you buy a discount package, you assemble bar and chain yourself and hope that the owner's manual is clearly written and comprehensive enough. Then if something doesn't function, you have to seek a franchised dealer—if you can find one within driving distance. If there's no franchised dealer with a repair shop, you'll have to strike up business with another repair shop.

On the other hand, you may be able to buy direct from a franchised dealer who also handles repairs. Just avoid a Saturday-morning visit. That's also the shopping day for other potential weekend loggers. Try to stop in when things are slow. Then a salesperson will take time for your questions and perhaps demonstrate a few models.

If you should decide to buy, you'll always know where to make a warranty claim or to get prompt repairs. Normally, the dealer will also agree to set up the saw (mount the bar and chain) in your presence and then fire it up to check carburetor adjustments and flow of chain oil. All the while you can be asking questions and taking notes. This is important, for every brand and model has its little idiosyncrasies that the mechanics know well. Here you can learn what kinds of mistakes most often bring chain saws back for repairs. One common problem, an embarrassing one for the owner, is that the saw failed to start because the owner forgot to flick on the ignition switch. Fortunately, there's usually no charge for this repair.

Before starting the saw at home, study the owner's manual well. Then begin your solo acquaintance with the saw. You've begun right, and you'll know whom to call if something goes wrong or if you simply have further questions.

WHAT TO LOOK FOR. If you visit several dealers, you'll likely get a look at more than several makes of saws. In this case, ask what types of saws the dealer rents out. He may be willing to apply a test rental toward purchase of a new saw, should you like the rental saw. But more important, if one brand predominates as a rental saw over others it may be because it's the hardiest one, needing fewest repairs, and then it may be the easiest to service. The dealer may admit this freely. That's a point in favor of the rental brand. Also find out which brand the dealer stocks parts for.

Size and length. Size and bar length are important. Most engines can be fitted with bars of several lengths. The range of bar options is based mainly on the engine's power, as well as balance. It's generally smart to avoid selecting either the smallest or the largest bar your saw can be fitted with. Using the longest bar possible tends to result in a slightly underpowered rig on very large logs. The shortest possible bar may leave you overpowered and regretting that you didn't save money by buying a smaller engine.

What bar length's best? As a rule it's best to use a bar that is longer than the diameter of most of the wood you'll be cutting. Here you'll usually have adequate power and you'll be able to cut through each log in one stroke, unless the log is stressed. And if you'll be felling trees, you'll find it easier to make simple wedge and backcuts rather than to fret with compound cuts. Yet excessively long bars have long chains with more teeth to sharpen, and longer replacement chains are costlier.

Anti-kickback devices. Kickback is the sudden, often violent upward or downward whipping action of the bar. This occurs when chain at the upper or lower curve at the nose of the bar is momentarily slowed or stopped by something solid. With the chain hung up for a fraction of a second, the engine wants to continue driving the chain around the bar, but the briefest of drags at the nose of the bar causes the engine torque to rotate the bar and engine themselves, kicking the bar up or down. This sort of kickback has accounted for about one-third of chain-saw-inflicted injuries. Operators have suffered severe cuts on their shoulders and backs when saws have kicked almost full circle. But the kick can sink the chain into feet, legs, arms, faces, and chests as well. So one important rule is to keep all body parts out of the plane a kicking saw could describe. Also, proper hand grip is essential. The accompanying photos illustrate these safety principles.

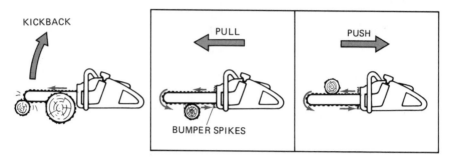

Most injury-producing kickbacks occur when the chain near the tip of the bar momentarily hangs up on something solid (left). Proper hand grip and a locked left elbow help lessen the kick. Violent pull or push can occur unless the saw's bumper spikes rest against the wood at the start of the cut. It's also important that the saw be fully revved just as the cut begins.

There are two main types of anti-kickback devices. The simplest is a safety tip on Homelite and Dynamark saws. It's just a hardened-steel fender that mounts over the nose of the bar. The tip prevents chain near the nose from hitting anything solid. Thus, no kickback. The second type of device was introduced by Stihl and now appears on a number of new makes. It consists of a chain-braking lever that stops the chain whenever the operator's front hand bumps a hand guard, as would often occur in an upward or backward kick of the bar.

Both the safety tip and the chain brake are desirable features, but neither is perfect. For example, the safety tip must be removed when you want to cut a log with a diameter greater than the straight edge of the chain. Without the tip, the saw is just as likely to kick as any other saw. Operators in a hurry sometimes find that mounting and removing the tip is too much bother, so they may operate for periods without protection.

The chain brake doesn't really prevent kickback. Rather, it is intended to halt the chain in milliseconds even though the bar and now-halted chain may kick far enough to contact you. Such a kick can still result in a bruise and a nasty laceration, from a halted chain, but this normally won't be of deep, bone-sawing magnitude. And remember, the front hand must bump the hand guard to engage the brake. This occurs only when the bar kicks up or back, not down, and when the front hand presses the hand guard forward. If the elbow of your front arm is bent at the moment of kick, the bar and moving chain could easily reach you before your hand bumped the hand guard.

In short, both types of anti-kickback devices are desirable, though not without shortcomings. The safest saw, with reference to kickback, would employ both the safety-tip option and the chain brake. But I know of no saw that offers both devices, and many saws offer neither.

If you have to make a choice, though, the chain brake offers advantages in addition to kickback protection. In some saws, it can be engaged routinely when you are starting the saw and any time you may violate a cardinal rule by walking

Homelite's Safe-T-Tip anti-kickback device mounts over the end of the bar by means of a single screw. But the tip must be removed before cuts can be made in logs with diameters exceeding the bar length. For a look at the chain-brake anti-kickback device, see the photo of the Stihl saw earlier in this chapter.

a few steps—however cautiously—with the saw running. The hand guard itself has another advantage.

Hand guards. The chain brake that employs a front hand guard also protects the front hand from the possible bone-slicing whip from a broken chain—another plus for the chain brake. The back hand is also vulnerable if the broken chain

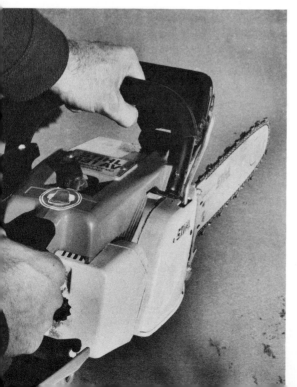

Proper hand grip includes opposable grip of thumb and fingers of both hands.

153

whips under the saw. Here a broad back hand guard can protect the knuckles. Some saws also employ catcher pins near the base of the bar that are designed to intercept a broken chain whipping under the saw.

Trigger interlock. This is a safety lever or button on the top of the back handle designed to lock the engine at idle speed. The interlock lets you hold the saw firmly while pulling the starter cord without fear that the engine will race enough to rotate the chain. Then the engine can't be revved until you press the interlock with your hand as you squeeze the trigger with your index finger.

Trigger interlock plus chain brake. Again, one cardinal rule is *never walk with a chain saw that's running.* But many people break the rule now and then. In this case, the trigger interlock and the chain brake in locked position provide some protection against unintentional chain rotation. Still, if the machine is out of adjustment the chain could move without warning.

Anti-vibration system. There's undoubtedly a correlation between the muscle-relaxing effects of vibrating chain saws and pharmaceutical vibrators. Though it's desirable to feel comfortable when operating a chain saw, it's not good to feel limp. Prolonged use of the vibration-prone chain saws can bring on limp-armed fatigue and a basically dull nervous system. That's a safety hazard.

Loggers who work long days with chain saws sometimes fall victim to "white fingers," or TVD (traumatic vasospastic disease). It's caused by vibration that cramps blood vessels in the hands and lower arms. The cramps may depart soon after a rest, and they may not recur until after prolonged use of a chain saw. But veteran loggers may suffer permanent damage. In this case, they can't work without pain, which may be especially severe if the hands are cold or wet.

Although TVD is normally an affliction of pros, it can begin to bother anyone who puts in long hours with a saw. The remedy is to work less with a saw, or to take frequent breaks, or to invest in a saw with an especially effective anti-vibration system. Most of these systems involve rubber cushions between the vibrating parts and the handles. They're worth the money if you cut more than a few cords a year.

OTHER FEATURES. Although the safety features described above should be high on your list of priorities, other features may be important to you too. Some of those below are useful, some are not.

Heated handles. Cold handles in winter can be uncomfortable at the very least, and they can aggravate TVD. Many small- and medium-size saws have rubber-insulated handles that are less cold than painted steel or aluminum, and they help reduce vibration as well. Also, gloves can freeze to bare handles, causing loss of control. Handles on some medium-size and large saws may be heated by electric rods or by the exhaust system. You may never need heated handles unless you work on cold days. But then they can be a blessing.

Electronic ignition. This sealed ignition system does away with a condenser and with points that are subject to wear and foulups. Timing remains set for the life of the saw. So it's goodbye to tuneups and new points.

Automatic sharpening. This feature adds more weight and expense as well as machinery that can be out of adjustment. Anyway, hand sharpening is easy and accurate if you use a clamp-on sharpening tool.

Proper cutting form includes locked left elbow (to minimize the effects of a surprise kickback), balanced stance, and saw chain in a plane that does not include the operator. Note also the protective goggles and the muff-type sound deadeners.

Electric starter. This is okay for garden tractors and other large engines, but if you can manhandle logs and hoist a chain saw, you can probably pull a starter cord. Why pay for frills?

BASIC SAFETY GUIDELINES. Before you do anything else, get familiar with your saw. Study the owner's manual carefully, and then reread it with the saw in front of you. Then develop the following good habits:

- Follow the manufacturer's instructions for fuel mix and for lubrication.
- Using gloves, check for proper chain tension before startup and periodically throughout the workday.
- Protect your eyes from wood splinters as well as grit by means of safety glasses, goggles, or a face shield that mounts on a hardhat.
- Protect your ears by means of well-fitting plugs or muff-type sound deadeners. Even if the saw's muffler reduces decibels from the engine to low, comfortable levels, the noise from the chain can easily reach an ear-damaging 90 to 100 decibels.
- Wear comfortable, though snug-fitting, clothing that can't get hung up in the work.

- Wear footgear that gives you best traction for the conditions. Steel-toed boots protect against saw cuts as well as crushing blows from logs.
- Start a gas saw in a well-ventilated area at least 10 feet from the fuel can and any spilled fuel. But first wipe all spilled fuel and oil from the saw. Otherwise, sparks from the muffler can ignite the gas, and the oil can make handles dangerously slick.
- Never smoke near a gas saw, and never operate it near sparks or flames.
- Before starting an electric saw, be sure the outlet protects you from shock by means of a ground fault interrupter (GFI). Be sure that cables and connections are in good condition.
- Start a gas saw on the ground, being sure that the chain will not come in contact with anything when you pull the cord. Here, one foot inside the back handle is standard for best control.
- Check to ensure that the chain and bar are receiving enough oil.
- Work only where you have good footing and where the saw won't bump obstructions.
- Use a proper two-handed grip at all times, being certain that the thumb and fingers of both your hands take a solid, opposable grip.
- To avoid kickback, keep the tip of the bar from touching anything inadvertently. Even so, always hold the saw so that an unexpected kickback won't whip the bar into any part of you. Here it's also important not to reach with the saw or crouch over it to get at wood. Work with your front elbow locked whenever possible.
- Never walk with the saw running. It's safest to carry a shut-off saw with its chain scabbard on. When that's not practical, normally you should aim the chain behind you. But when walking downhill with an exposed chain, aim it in front so you won't land on it in a slip and so you can get rid of the saw if necessary.
- Let the chain hit full speed just as it enters the wood. Early revving causes the engine to race too much. Late revving will cause binding and clutch slippage. For best control, have the bumper spikes in contact with the wood as the chain begins cutting.
- Touch up the chain teeth every couple of hours, or whenever the sawing starts producing sawdust rather than chips of uniform size.
- Never let a helper approach within range of the saw or into a 15-foot radius of you when you are bucking small-diameter wood. A saw can kick out a small piece of bucked wood at high velocity.
- Don't work when you are fatigued. Take breaks often.
- Don't get overconfident.
- Perform all maintenance on the saw before you put it away for the day.
- Check additional guidelines for felling and limbing in Chapter 17.

16 | Hand and Power Splitting

THE NEED FOR splitting depends on the type of wood, the diameter of the log, and the time available for seasoning. The idea is to reduce the wood's moisture content to a low level. Some barks retain moisture more stubbornly than others. Large logs may never release much moisture from their inner layers unless they are split. So when there is only a month or two from cutting to burning, most green woods over 3 inches in diameter should be split. But when there's 6 months of balmy weather for drying, most logs up to 4 and even 5 inches can be left unsplit.

Years ago, all splitting was done by hand. Today you can also use a power splitter. These machines can split eight to sixteen cords of wood per day, if you can keep up with them. If you are splitting for pay or starting a firewood business, a power splitter is a good investment. But if you are healthy and plan to split less than six cords a year, a power splitter may cost you more money and maintenance time than it's worth. Besides, power splitters consume gas and oil and emit noxious fumes, and their noise level depends on the quality of the muffler. These machines mask the pleasant fragrances of freshly split woods. They inhibit conversation. As for your pausing to appreciate a wood's grain or to watch a squirrel, forget all that, for you'd be wasting gas.

Hand splitting need not be so arduous that it makes you huff and puff. It can be an easy, rhythmical form of recreation—evenly spaced whacks counterpointed by bird songs, fresh air, and quiet.

TOOLS FOR HAND SPLITTING. The tools you'll need depend on the kinds of wood you split and the wood lengths. If you're firing a stove that takes lengths of only 12 to 14 inches, you may be able to get by with a hefty ax. In this case, it's wise to obtain fairly easy-splitting woods whose large knots and crotches have been crosscut with a saw. Otherwise, knots and crotches can make you cuss for want of more adequate tools. These may include steel wedges, and either a sledge or a splitting maul or both.

Ax. An ax for splitting should be single-bladed, with a 3½-to-5-pound head. Its edge should carry a broader bevel than an edge used for limbing and felling. It should be sharp enough to cut through interlocking grain but not so sharp and thin that a hard knot can shiver it. The head should be broad at the heel rather than narrowing so that it won't bind when sunk deep into an uncooperative log. The handle, or helve, should be of hickory about 36 inches long. By way of comparison, the well-known Hudson Bay style of ax generally has a head weighing 2½ pounds that is fitted with a handle about 26 inches long.

Sledge and wedges. These are designed for splitting jobs too tough for an ax alone. Never drive a sledge onto the heel of an ax to force the ax deeper into

Shown left to right are a 6-pound splitting wedge, an 8-pound splitting maul, a 5-pound wedge, a 1½-pound hatchet, and a 5-pound ax. Perhaps the most treacherous tool in the lot is the little hatchet.

a log. An ax head is not built to withstand heavy blows on the heel. Yet many people do whack ax heads and often wind up with a mushroomed heel and a sprung eye (the hole that receives the handle). Any sledge of 6 pounds or more and a handle of over 30 inches will do.

Splitting wedges are made of steel and weigh 4 to 6 pounds. (The plastic and aluminum wedges sold in hardware stores won't survive the rigors of splitting. They are meant to hold chain-saw kerfs open. The soft material of these wedges won't damage a chain-saw blade if the teeth of the chain accidentally strike the wedge.)

It's important to have two splitting wedges rather than just one. The second wedge is used to free the first. If, instead, you customarily try to hammer a lone wedge through knots, you'll wind up mushrooming it after a short time. And a single wedge can become hopelessly lodged in a log.

It's a good practice to tap the first wedge only until it stops advancing easily. Then start the second wedge. This may be enough to split the log. If not, the second wedge will spread the wood enough for you to drive the first wedge easily again or reposition it for better effect.

Splitting maul (hammer). Mauls are essentially hybrids of axes and Oregon-style wedges. They have a blade on one end and a combination hammer-wedge head on the other. Head weights range from 6 up to 12 pounds. Blades are only 3½ to 4 inches wide, compared to 4½ to 5 for a lighter ax. But the splitting hammer is broader across—2 to 3 inches across. This added breadth, together with the added weight, results in blows that deliver increased momentum and splitting force. The hammer end of the head can serve as a wedge head that can be struck with a sledge, or it can be used to drive regular wedges.

In my opinion, the splitting maul is the most versatile and efficient of all splitting tools. I use an 8-pounder for almost all splitting tasks. When it's not enough alone, I either drive it farther with a sledge or switch to a sledge and wedges.

Hatchet. A 1½- to 2-pound hatchet is handy for splitting kindling to finger size. But it's safe to use only as long as the kindling splits easily. A hatchet becomes a treacherous splitting tool as soon as you start trying to deliver hard blows with it. More on this a bit later.

Wooden splitting block. Blocks are essential because they provide a solid, unyielding surface that sandwiches the wood to be split between the block itself and the tools. Without a block, the splitting wood must be placed directly on the ground. Here the ground :   es like a cushion, dampening the impact of the blows.

Woods notoriously tough to split (such as the elms, beech, and the crotch sections of most hardwoods) make wonderful splitting blocks. They can take many direct blows and absorb tons of shocks sent through the splitting wood without showing fault lines themselves.

The best block size depends on the job. In diameter, it should be at least double that of *most* of the wood being split. This ratio has several values. First, it gives a broader base for the wood to jump around on without falling off. Wood will jump around with impacts, and you'll find the stuff far easier to

For strongest handles, grain should run essentially with the cross-sectional long axis.

Stringy cross grain of American elm holds even though a wedge and maul have spread the top end of log 2½ inches and the bottom nearly an inch. Elm's resistance to splitting makes it a long-lived splitting block. Bark should be stripped off immediately to prevent breeding of elm bark beetles.

center again than to pick up from the ground for another blow. Second, the broader block stands a better chance of catching an errant blade before it strikes the ground. Third, a broad block often catches one or both pieces of the split wood before they tumble to the ground; if these same pieces need to be split again the convenient lie becomes a blessing. After some practice, though, you'll often be able to adjust the impact so that the wood doesn't break apart until you administer a calculated twist that makes the pieces fall where you want them.

Block height should depend on the length of wood being split. The idea is to use a block that lets you make good initial impact at about your mid-thigh level. This way your back is still about vertical on impact and the handle of the tool is nearly horizontal. The value here is less one of might than of control.

This same block provides convenient working height when you are starting wedges, using a shortened grip on the hammer. The less stooping you have to do, the more efficiently you'll perform.

HOW TO BUY SPLITTING TOOLS. Find a hardware store that seems to have the favor of local craftsmen. Here axes, sledges, wedges, and mauls should have a solid, quality look to them. Hickory handles won't be painted to hide wood grain. The bottom cross section of the handles should show grain running parallel to the long axis—not perpendicular or diagonal—for best strength. Prices are likely to be higher than for similar tools in discount centers, but cheaper prices in tools usually mean lower quality.

TECHNIQUES OF HAND SPLITTING. The most unwelcome weekend guest is the fellow who progressed from Boy Scout hatchet to home-run king and then to your woodpile. This fellow is strong, well coordinated, and accustomed to making objects yield. He will likely place each log section on the block and de-

160

liver a mighty blow to the core of each. If the first few pieces are straight-grained ash and maple, the fellow will be popping them apart with ease. If the fellow knows he has spectators, the display of prowess will continue—only until a hefty elm log appears. Here a blow to the core may produce only a thud and bounce of the tool on the surface. If you let the fellow have his head, he may be powerful enough to break the tool handle.

Woods of different tree species have reasonably predictable splitting qualities. Most woods are easier to split green than seasoned, and frozen logs often split readily. Yet parts of an individual tree may pose widely different splitting challenges. Straight-grained sections tend to be easiest. Irregularities such as knots and burls tend to be balky. Wavy grain and arcing grain, most common where the trunk joints the roots, can be difficult. Birches and some oaks have a lot of wavy grain. And elms, especially, have interlocking fibers that may see you bury two wedges and force the split several inches apart with the log resolutely refusing to fall asunder. Crotch bases can be almost impossible to split; they are easy to recognize because the annual rings in them emanate from two cores, forming a figure-eight configuration.

Efficient hand splitting calls for as much thought as muscle. If you use a maul of about 8 pounds, the falling weight of its head may often generate enough momentum to accomplish the split. In this case, the main physical effort occurs during the *upswing*. Then during the downswing, the arms and hands merely provide firm guidance to the point of aim. The lighter the splitting tool, the more effort and speed are required during the downswing. Here, wielding a lightweight ax, you may have to get downright savage to keep pace with someone swinging a maul.

Once you become accustomed to your splitting tool and the types of wood in your woodpile, you'll be able to gauge in advance how much force will be needed for each split. Sometimes the weight of the maul will be sufficient. Sometimes the maul needs a little added "starter" at the top of the swing. Rarely are downright savage blows called for. Savage blows soon fatigue both you and your tool. And with your own fatigue comes diminished concentration and control. I've no statistics to prove it, but I'd bet that most splitting accidents occur after the victim has been swinging too hard for too long. Glancing blows and tired hands can bring disaster. Or, on a less dramatic note, overshooting your point of aim can make the tool handle strike the work. When that happens, you know you've gotten too casual.

Handles cost about $5 and take an hour or two to replace, depending on the amount of difficulty you have removing the stub of the old handle. Good replacement handles come with metal wedges and little instruction sheets.

One alternative to savage blows includes wedges and either a sledge or the hammer head of a maul. Here it's necessary to stand close to the work and take a short grip on the sledge or maul. To split the log in half, place the cutting edge of the wedge in a crack or on a medullary ray beyond the center of the log, as shown in the accompanying drawing. (In live trees these medullary rays connect inner layers of the tree with the bark for storage and transfer of food, and they conveniently reveal planes of weakness that you can exploit for splitting.) Then tap the first wedge in far enough so that you can step back to deliver harder blows. But again, easy does it. When the first wedge no longer

Using an easy stroke and deft handwork upon impact, you can split segments off a log without knocking it off the block. Twigs scattered over snow improve footing.

drives easily, start the second wedge along a near-side medullary ray so that the split will bisect the log. Here the first wedge remains conveniently out of the way, so you won't have to change positions in order to avoid hitting it. Using the two wedges, you can either drive them both straight through or use one to free the other for repositioning. Tough-splitting woods have been known to bury three wedges without carrying the split.

"Daisying" is a popular alternative to savage blows, and it may or may not require wedges. It's best used on logs over 8 inches in diameter. As shown in the accompanying drawing, daisying involves splitting off arcs (like petals of a daisy) around the circumference of the log first, and then splitting triangular sections from the remaining polygon. From the drawing, you might think daisying lets you walk sidestep around the log, lopping off a section with each step. In practice, though, the small piece falls off like a daisy petal, all right, but the parent piece often twists or falls off the other side of the chopping block. Anyway, it's not practical or safe to walk around a chopping block, owing to the small mountains of split pieces to hurdle and stumble through. So daisying usually involves repeated turning of the parent piece for the next blow, or stepping one step to right or left, or setting the tool down and lifting the fallen parent piece onto the block again. In all, the technique is a challenging one because it makes you give as much effort to the resulting lie as it does to the split. But it helps you whittle a large log down to stove size mighty fast.

KINDLING. Unless you can keep a fire going continuously through the heating season, you'll need to put some kindling by. The amount you'll need de-

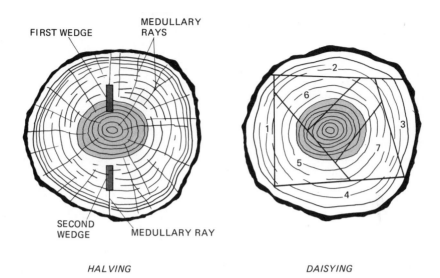

FIRST WEDGE

MEDULLARY RAYS

SECOND WEDGE

MEDULLARY RAY

HALVING

2
6
1
3
7
5
4

DAISYING

Here are two good ways of splitting large logs. To halve a large log, start the first wedge in a medullary ray or a crack on the far side of the log. When the first wedge no longer drives easily, start the second wedge. Daisying usually goes faster. Here you simply work around the outside, splitting off sections like daisy petals.

A small splitting block inside a wheelbarrow puts the work at a comfortable height, and the wheelbarrow catches most of the pieces for easy carting. In most cases, it's safest to make the wood and the blade descend together, as shown in the photo, pulling the lower hand away at the moment of impact. The alternative is to hold the wood on the block with one hand and strike with the other. But that leaves the lower hand a stationary target for a misdirected or glancing blow.

pends on the number of times you expect the fire to die. For most of us, the fire dies more often than we expect.

Prime kindling should be dry wood of finger thickness. Kindling of slightly larger diameter can be used for the transition to large pieces. Many people favor softwoods here because their resins ignite at lower temperatures and because their resins yield higher Btu per pound of wood. But seasoned hardwoods serve well. Hardwoods weighing more by volume than other hardwoods yield more Btu before they fade. I favor woods that split easily with a hatchet.

Some people hate to split kindling because the task either makes them work bent over or crouched down. One authority recommends dropping to one knee, a position that's punishing on rocky, wet, or icy ground.

For comfort and best control, the work should allow you to stand and strike wood at waist level. Few chopping blocks are tall and stable enough for this. But a hefty wheelbarrow with a small-diameter block inside is ideal. Just place the block inside so that split pieces land in the bin. Pieces in the bin that are still too large for kindling are within easy reach for another split. Then with the wood heaped around the block, remove the block, retrieve the few pieces that have fallen to the ground, and wheel the binful to storage.

Hatchets. Again, these things can be treacherous. They must be controlled with one hand, and they impact close to your free hand and your legs. Splitting with a hatchet is sensible only as long as the wood splits easily. If you encounter a bulky piece of wood, switch to a chopping block on the ground and a maul or heavy ax, gripping the tool with one hand near the head. The heavy tool can deliver increased splitting force even though it seems to descend in slow motion, compared to the lightning blow needed from a hatchet on the same wood. Then too, the inertia of the heavier head keeps it from glancing off as a hatchet head can. And finally, the heavy head is descending slowly enough to allow you time to halt the action in the event you really miss your aim. It's possible to do an efficient job of splitting kindling using only a maul. Still, a hatchet allows faster, more rhythmical work on easy-splitting wood.

POWER SPLITTERS. An expert with a splitting maul and wedges can keep up with a power splitter for an hour or two if the logs are short and straight-grained. Here, time is lost mainly in lifting and placing wood for the split, rather than in the splitting motions themselves. But some power splitters will easily outsplit a man when the logs are long and the wood is bulky.

There are basically two types of power splitters: the hydraulic-wedge type and the screw type. Both require gasoline engines.

Hydraulic splitters. These units employ a hydraulic cylinder that forces the log end against a large stationary wedge. Heavy-duty models generate 10-ton splitting forces and can handle long logs. Some models can be coupled to the hydraulics of farm tractors. Others are driven by their own 4- to 8-horsepower engines.

Screw-type splitters. These use a threaded cone that screws itself into a log, the widening base of the cone prying the log apart. Some units connect to the power take-off (PTO) of tractors. Others can be bolted to the rear hub of a car, truck, or garden tractor. Still others are driven by their own engines.

A GALLERY OF POWER SPLITTERS

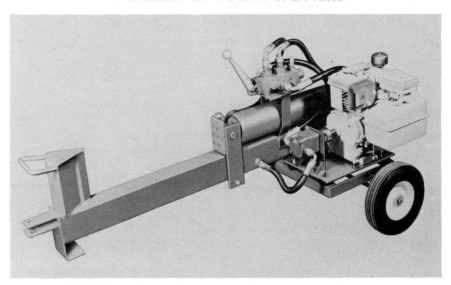

The Mighty Mac LS 185 by Amerind-MacKissic accepts logs up to 19 inches long. It is just 63 inches long and weighs 283 pounds. Its 5-hp Briggs and Stratton 4-cycle engine can deliver 22,000 pounds of hydraulic splitting force. Total cycle time: 17 seconds. Amerind-MacKissic also offers a 7-hp unit that takes logs up to 26 inches long.

Self-powered Nortech "Screw Wedge" log splitters are available with 8-hp engine and 5-inch-diameter screw wedge and with 9-hp engine and 6-inch-diameter screw wedge. Nortech also makes units that run off farm-tractor and Gravely-tractor PTOs.

This Stickler by Taos Equipment Manufacturers is driven by a tractor PTO, and it mounts on a three-point hitch. Taos also makes a car-mounted Stickler.

This combination hydraulic splitter and cordwood saw is made by Richard Pokrandt Manufacturing. The splitter is powered by tractor hydraulics, and the saw is powered by the PTO. Splitting force and speed depend on the tractor's hydraulic system. The saw blade is 30 inches in diameter. Saw and splitter weigh about 300 pounds each.

Price ranges. Hydraulic splitters cost from $600 to over $1000. Screw-type splitters range from $200 to over $800. For each type, lower-priced models rely on drive systems of other vehicles and higher-priced models are generally self-powered.

Self-powered units. Aside from their higher cost, self-powered units give you yet another engine to maintain and service. This is a pain to many people but an exciting challenge to others. On the plus side, self-powered units can be wheeled to work sites in almost any terrain. Smaller models can be transported in a pickup bed. Large models can be towed.

Hydraulic types: pluses and minuses. These units are often the choice of pros because they can tackle any log, regardless of diameter and wood type. In this case, the wedge slices through all interlocking wood fibers as the log is forced past. Some units have cycling times of only 12 seconds. Others need 30 seconds. Logs tend to shift as the wedge begins to follow grain, so sometimes you must stop the machine to reposition the log. Most designs require that you lift the logs onto a splitting bed, and lifting is real work.

Screw types: pluses and minuses. Screw-type units perform best on straight-grained, moderately easy-splitting wood. They'll split difficult woods with interlocking fibers, but not as cleanly as wedges do. Here you may wind up with a log split as wide as the widest part of the screw. Yet the split may not have carried through the log. Or, since the screw really pries the wood apart, it may leave many stringy wood strips on the face that have to be trimmed with an ax before the wood will stack neatly. One company frankly warns that its splitter should not be used on tough-splitting sweet gum and black gum found in the South and the East.

Yet, screw-type splitters can be dependable workhorses as long as you cut tougher-splitting woods short enough so the split will carry the length of the log. Homeowner models are rated for lighter duty than high-priced pro-type models. One big advantage that most screw-types have over hydraulics is that the log need only be tipped into position rather than lifted.

This Unicorn by the Thackery Company bolts to the rear wheel of a vehicle. With the wheel jacked up and the car idling in second gear, the screw's coarse threads work at low rpm. Note the emergency shutoff switch mounted on the side of the vehicle. Unicorn models are available for cars, trucks, garden tractors, and tractor PTOs. Prices range near $300.

17 | Doing Your Own Logging and Tree Work

BIG-TIMBER LUMBERJACKS concern themselves with felling trees so there's minimal loss of potential lumber. But if you're just doing a little tree work around the neighborhood, you don't have to worry about damaging the tree at all. All you have to worry about is dropping the tree onto power lines, or your neighbor's house, or his family, or his outbuildings, or his septic field, or his fences, or his pets, or the people next door, or neighborhood kids who've come to watch. Then again, the tree could land on you. Or your chain saw could hit a nail grown over by tree bark. Anyway, if you agree to fell an old neighborhood tree, you'll have a bit to think about—even if the wind isn't blowing.

It's so tempting to do tree work in the neighborhood. The wood needs only a short haul. You're doing the neighbor a favor. You'll get some fresh air and exercise. And you may be paid well. Besides, every tree of 20-inch trunk diameter represents about a cord of firewood to you. If you are desk-bound all week you may even fantasize about chucking the old routine and launching a tree-removal business. We all have our fantasies.

Let's get realistic. You may need a license. When you contract to remove trees, you are liable for any damage you do to property and persons. You may be maimed or killed. The insurance organizations advise that most homeowner policies cover a person merely helping a neighbor and collecting the firewood. But a pro, even if his compensation is small, should carry a special comprehensive liability policy with a minimum of $100,000 coverage.

FOREST AND WOODLOT LOGGING. If you live near a state or federal forest, you may be able to remove cull trees designated by foresters. Normally you pay less than $5 per cord plus about $2 for insurance. This approach can give you more than low-cost wood. If the forester accompanies you to the cutting site, he may be willing to explain local tree-management principles.

In some cases, foresters may want to clear-cut level terrain where there's no danger of erosion to provide habitat for deer and smaller animals. In others, only selective cutting may be employed. You can learn a good deal about selective cutting by analyzing the relationships of cull trees to the protected trees. Normally, the protected trees are crop trees, marketable for lumber. Others, such as nut trees, may be protected because of their importance to wildlife, regardless of lumber value. Or pure stands of a single type of tree may be thinned and the openings replanted to provide a mixture of trees less susceptible to the ravages of individual blights or insect invasions.

Many of the cull trees will be diseased, crooked, or otherwise unsalable. Others will look good themselves but will be competing with more desirable trees, perhaps even threatening to shade them to death. The most valuable crop trees will normally be straight-trunked, with only small branches, if any, along the lower portion of the tree. Many of these branches should show signs of self-pruning. Crowns of these crop trees deserve free space on at least a couple of sides. The crop trees will usually be spaced about 15 to 25 feet apart. Under such management practices, about 100 crop trees with 20-inch trunks may result from an acre of land that began with 4000 to 6000 sprouts and trees of all descriptions. Meanwhile, small sprouts and cull trees will provide a sustained yield of firewood—perhaps as much as a cord per acre per year, forever.

If there are no state or federal lands nearby, you may decide to buy your own woodlot or help a landowner manage his. In scouting for acreage, remember that best burning woods tend to grow on well-drained soils. You should also consider how you will remove the logs. Here roads for vehicle access become an asset. Hillside acreage may let you skid logs down to roadside over the snow. But before you bid on the land, make a walking survey of the kinds of trees. If you already know how to identify trees in your locale, you'll be in good shape. If not, you may be able to engage the help of a forester or logger. A good tree identification guide or two will prove valuable too.

If you intend to buy land, find out first what your taxes will be. In some areas taxes are based on the market value of the land. So if you pay more than the old assessed valuation, the taxes may immediately escalate proportionately. If you prefer to log someone else's land, you will normally have to pay "stumpage," which may be based on the number and size (scale) of trees you fell or the number of cords you remove.

Either way, few woodlot managers expect to get rich. Yet a woodlot offers boundless opportunity for learning and exercise. Dollar returns will depend mainly on the kinds of saw logs you can haul to a sawmill, and this requires costly equipment or the fees of truckers with hydraulic log loaders. Many small-scale woodlot managers feel justly rewarded if they can meet their costs each year. For them the woodlot provides rewards other than money.

Woodlots managed solely for dollar yield often look like parks. But this is not environmentally sound. By leaving some dead and dying trees and tree tops, you provide breeding grounds for creepers and crawlers that are vital links in a forest's life chain. Large, heart-rotted trees can provide nests for birds and small mammals. Downed branches, themselves, should not be removed because they contain most of the nutrients the trees draw from the soil. By leaving the tops you allow the branches to decompose and refortify the soil. If you make brush piles, you will be providing safe harbors for small mammals, and if you do most of your felling in the fall, the tops will provide browse for deer and rabbits over the winter. The point is, tree cutting can provide more than mere lumber and fuel.

"LOGGING" WITHOUT WOODLOTS.

If woodlots and forests are too far away from your home to make them practical fuel sources, your range of other sources may nevertheless be great. This may include logging without felling.

Dumps and landfills. These made easy pickings before the fuel crisis of the early 1970s. But now very little burnable wood is discarded. Access to dumps and landfills is often by permit, and you may be required to pay for wood you remove. Besides, these logs are often covered with mud and grit that will dull saw teeth.

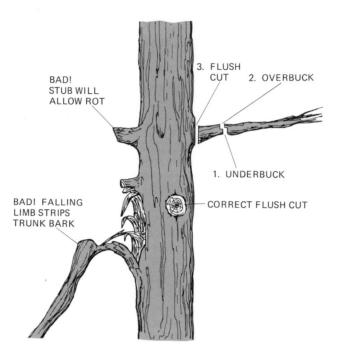

BAD!
STUB WILL
ALLOW ROT

3. FLUSH
CUT

2. OVERBUCK

1. UNDERBUCK

BADI FALLING
LIMB STRIPS
TRUNK BARK

CORRECT FLUSH CUT

Pruning cuts on the left-hand side of the tree were made incorrectly. Those on the right are correct.

Orchards. Fruitwoods have high Btu values. Many are pleasantly aromatic as they burn. Years ago, orchardists welcomed people willing to glean the prunings. Today, it's hard to find an orchardist unacquainted with the value of his prunings. So to obtain prunings you may have to pay for them, unless an orchardist owes you a favor. Otherwise, your chances of gathering goodly quantities of fruitwood are best if you plant your own orchard.

Following road and line crews. Line crews for power companies and road crews for municipalities and highway departments are good people to tail. Crewmen often carry home enough wood for their own needs throughout the year and welcome anyone who arrives in their wake with a truck. They may even cut the wood to lengths you specify. And your own loading of the wood saves the crewmen some backbreaking labor. Sometimes schedules of roadside tree removals are announced in newspapers. Otherwise, you may have to phone officials for information. If you can establish an "in" with a foreman or a crew member, you'll get opportunities.

Storm cleanup. Storms and freezing rains often knock down large tree limbs. If you happen upon them along a roadside right-of-way, the limbs become yours by virtue of your promptness. Be sure your vehicle is parked safely off the road, though. And don't mess with limbs resting on power lines. Storms may also knock down residential trees. If you arrive at a residence before the full-time tree contractors, you may receive both the wood and a payment for your efforts.

Carrying a saw, always. It's smart to carry a bow saw, at least, in your vehicle. Then you are prepared to section and load limbs that commonly lie along secondary roads. You may also find work clothes and boots handy, should you encounter a real windfall. A sledge and wedges will let you split large logs to loadable size. You may surprise yourself at the amount of wood you can pick up over the course of your regular travels.

Farm edge. Busy farmers may welcome your offer to thin the tree lines along field edges and fences. Tote roads may allow close vehicle access, or you may have to wait until fields are frozen so you can drive straight to work sites. You may have to pay stumpage, or you may be able to negotiate a wood-sharing agreement.

Pruning. I know a roofing contractor who acquires many cords of wood simply by pruning limbs growing too close to the roofs he covers. But pruning above roofs can be tricky. A limb need not be terribly large and it need not fall far to damage shingles and roof boards.

Hardwoods should be pruned when they are dormant. That's between late autumn and early spring—after the leaves have fallen and before the sap begins to flow. Conifers should be pruned in late spring or early summer, before they begin to grow rapidly.

Accidentally stripped bark below the pruning cut will leave the tree vulnerable to disease. So starter cuts should always be made on the bottom side of the limb. That way, when the second cut, from the top, severs the limb, the limb won't tear off trunk bark. Always cut limbs as nearly flush with the trunk as possible. Otherwise stubs will fail to heal properly and the exposed wood will allow rot to enter the trunk. Dress all cuts over ½ inch in diameter with tree-wound dressing.

FELLING AND BUCKING RESIDENTIAL TREES. Again, this requires a license in some localities, and if you work for pay you should carry comprehensive insurance. At the very least, you should first acquire extensive experience bucking, limbing, and felling in the woods before you take on residential trees that may be "over your head" in many ways. The dangers here to life and property can be great.

Often the chief dilemma is lack of felling room. Homeowners themselves will often fell trees that can be dropped in one shot, so if you are asked to bid on a residential job, you'll normally be faced with medium-size trees that must be topped, or large trees that must be lowered limb by limb before you can dump the trunk. This always involves climbing and usually requires adroit use of ropes. On some trees, you may be asked to remove only limbs that shade a house or threaten disaster.

It's easy to underestimate the challenges of residential topping and limbing. From the ground, limbs only wrist-size at the butts appear to be easy drops. When severed, these small limbs are light enough to be supported in one hand. Yet when they free-fall 50 feet, their leaves and twigs serve like tail fins on a missile, guiding the butt earthward as the falling mass builds velocity and momentum. Such a limb may plunge 8 to 12 inches into a lawn, and it will break through frozen turf as though it were crackers. I once accidentally dropped a small limb from a 50-foot perch. In memory now, the limb descends in slow motion toward a wooden staircase. In fact, the limb fell fast, because it wiped out a 2×8 step as though it weren't there.

Need for limbing. You may decide to limb a large tree simply because there is no felling room in any direction. Or you may want to limb on one side so the remaining limbs weight the tree in the direction you want it to fall. Either way, you'll be in for some climbing, as well as aerial engineering that exploits the powers of gravity, the swing of pendulums, and the mechanical advantages of fulcrums, pulleys, and ropes.

The accompanying drawings show solutions to a few of the challenges that are common in residential felling. Of course, there's no single best way for every situation. Each tree presents unique combinations of problems that arise from its size, shape, and condition, as well as the amount of room available for drops.

Equipment. It's important to begin a tree job adequately equipped. Without the right equipment you may endanger yourself or damage property, or you may find you must abort the job midway and ask a fully equipped contractor to finish up for you.

One way around trouble is to bid only on part of a job. To do this, you can bid on limbing and removing the tree after it's been topped or felled by a full-time contractor. Just tell your prospective client that he should ask a couple of contractors to bid on both the topping or felling and the removal. If your bid on removal is less than that of the contractors, you'll likely get the wood you want plus a few bucks that you can apply toward more equipment.

Saws. If you're in no hurry, a couple of bow saws can handle trees with diameters up to 20 inches. A 42-inch bow saw will give you two-man capability for felling and for sawing the large trunk sections. The only obstacle here is the 11-inch clearance from saw blade to bow. When the bow hits the bark you'll either have to move the saw to the other side of the log or else roll the log. A large bow saw is a workhorse that one man can wield easily on wood of all diameters, as long as the wood doesn't begin to move with the blade. Then it's often better to switch to a 30-inch bow saw that leaves one hand free to steady the limbs.

An old-style two-man crosscut saw has advantages over the big bow saw if you are making a deep back cut in felling. Here you can saw 8 to 10 inches in and then start a wedge or two in the kerf to keep the tree from "sitting back" on the saw and to lift the tree toward the intended direction of fall.

Since the penetration of the face cut, or notch, need be only about one-third the diameter of the tree, a big bow saw has adequate clearance to make both top and bottom face cuts on trees of 20-inch diameter. Then too, you may prefer to chop the notch out, after you've used a saw for the horizonal cut.

Avoid using a chain saw for felling large trees until you've become skilled with it in bucking and in felling small trees. Felling cuts require that you hold the saw at awkward angles sometimes, unless you compromise safety by placing yourself in the saw's kickback zone. That's bad. Chain saws are especially risky to use for felling when the bar is much shorter than the trunk diameter.

(Text continued on page 176.)

Use a plum line to check for natural lean of a tree. Otherwise, illusions created by sloping terrain or other leaning trees can lead to unfortunate felling cuts. The plumb line also helps you check for symmetry. In some cases, a trunk leaning slightly one way may be counterbalanced by limbs on the other side. Always plumb from several positions at least 90 degrees apart to confirm.

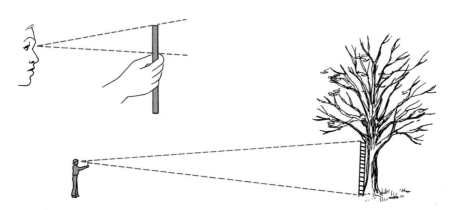

Prior to felling most trees, you'll want to determine height. Then you'll know for sure how far the tree will reach when it falls. One of the simplest methods, as shown, involves visually telescoping a known increment up a tree. To do this, place a ladder or pole of known height at the base of the tree. Then walk back two tree lengths or more. Holding a small stick at arm's length, align the top of the stick with the top of the ladder, and align the end of your thumb with the bottom of the ladder. Then visually telescope the ladder's length up the tree. The number of increments times the ladder's length gives you the tree's height. Perform this a couple of times to confirm.

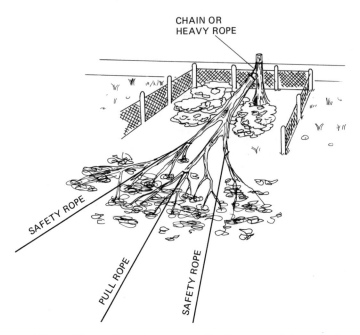

CHAIN OR
HEAVY ROPE

SAFETY ROPE

PULL ROPE

SAFETY ROPE

You can fell small to medium-size trees so the top drops into a small opening, while the butt is held by chain or heavy rope secured like a pair of handcuffs. Then you can make quick work of the top with lopping shears and a bow saw.

This large tree (top right) cannot be felled whole in any direction. But limbs on the left-hand side can be cut and swung to the right if the large right-hand limb is used as a boom for the rope. The ground man uses the guide rope to prevent the limbs from swinging back toward the house and to maneuver the limbs as he lets the support rope slide through his other hand (bottom right). It's vital that both men know how large a limb the ground man can handle in each situation. (Good judgment is more important than strength.) Later, the standing trunk can be dropped onto limbs laid to protect the lawn and any underground pipes. The trunk should be secured with safety line, top and bottom, to be sure it doesn't hit the house or steps. These lines can be run from other trees or from heavy vehicles.

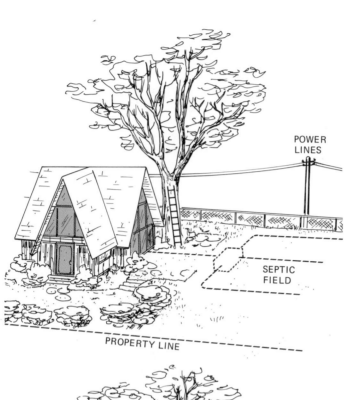

POWER
LINES

SEPTIC
FIELD

PROPERTY LINE

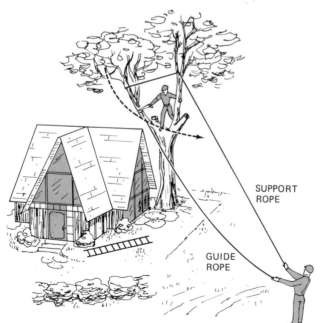

SUPPORT
ROPE

GUIDE
ROPE

175

When the bar is too short, you will need multiple cuts that are difficult to make with precision. The goal is usually a classic section of hinge wood to hold the tree on course as it falls. The accompanying drawings show some basics.

There are advantages to completing a back cut with a bow saw or two-man crosscut, even though you use a chain saw to make most of the back cut. The hand saw will give you a straighter edge on the hinge wood than a short chain-saw bar can, and the hand saw lets you hear warning cracks should they occur before you expect them. Then it's time to either drive the wedges deeper or draw up the pull ropes, or both.

Saw wedges. When used in chain-saw kerfs, wedges should be of soft material that won't damage teeth that accidentally nick them. Saw wedges are made of aluminum, magnesium, plastic, or wood. Wedges are used in bucking

Here the top half of the conifer nearer the house has been rigged to the next tree so that the top will arc away from the house. If the second tree is to be saved, its trunk should be padded with blankets that are held in place by bound vertical poles. The pull rope should be anchored and tensioned with a come-along or a couple of stout fellows. Later, the bottom section of the first tree can be felled using similar rigging. But first the felled top section should be laid over the walkway to protect it.

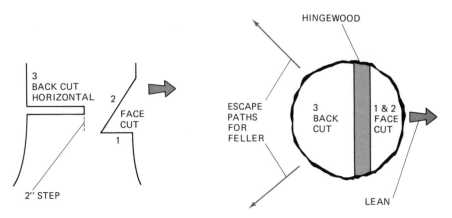

Here are conventional cuts for felling a tree in the direction of its lean. This lean may be natural, or you may create it by means of limbing, or by driving wedges into the back cut, or by use of pull ropes. The face cut is normally about one-third the trunk diameter. The back cut should be horizontal, ending about 2 inches higher than the intersection of the two facing cuts. This provides a step that prevents the tree from jumping back off the stump as it falls. Hinge wood itself should be about 2 inches wide for large trees, decreasing to about 1 inch for small trees. Note that the hinge-wood cuts are parallel and lie perpendicular to the intended direction of fall.

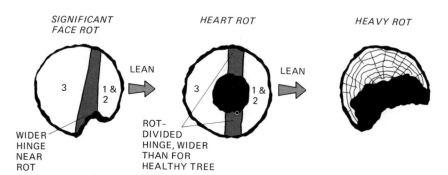

Trunks of large trees should be sounded for rot by hitting them with a sledge or the heel of an ax. Hollow sounds or mush give reason for pause. Since rotted wood has little holding value, felling cuts should be planned as though the rotted wood were cut away already. Trunks with only slight amounts of rot can be topped to reduce stresses on the trunk when it is felled. But don't climb a heavily rotted tree. Leave it for a well-equipped contractor to pull down or else top from the vantage of a hydraulically lifted cherry picker.

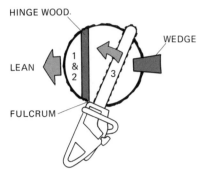

HINGE WOOD.

WEDGE

LEAN

FULCRUM

When a chain-saw bar is longer than the trunk diameter, back-cut as follows: Place the saw spikes (sometimes called bumpers or dogs) on the trunk so that you can rotate the bar, using the spikes as a fulcrum. Make one fanlike back cut to form the back of the intended hinge. Never cut to the notch.

When the bar is shorter than the trunk diameter, use the back-cut method shown. This provides a good view of the work while allowing a reasonably safe body position as well as good control. After starting the second fanlike back cut, remove saw and start a wedge.

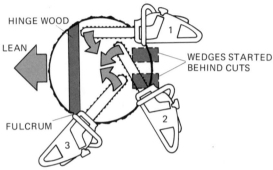

HINGE WOOD

LEAN

WEDGES STARTED BEHIND CUTS

FULCRUM

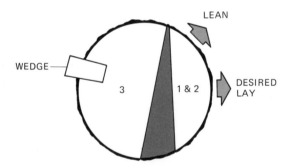

LEAN

WEDGE

DESIRED LAY

A hinge that is wider at one end will tend to bring a falling tree toward the wider end. This cut may be used to correct somewhat for a lean slightly away from the desired lay. Caution: This is a risky method in residential areas. Surer means include limbing and wedging.

A trunk with excessive lean will tend to break off and split vertically during the back cut because wood fibers at the face are highly compressed and at the back cut highly tensioned. The result is a "barber's chair." You may first be able to reduce the amount of lean by topping. If lean remains excessive, a small face cut and two shallow side cuts, as shown, should precede cut 4, which should proceed rapidly to the intended hinge.

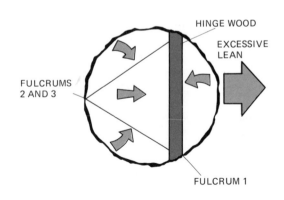

HINGE WOOD

EXCESSIVE LEAN

FULCRUMS 2 AND 3

FULCRUM 1

178

to prevent a log under tension from pinching the blade, and they are used in felling to lift the tree in the direction you want it to fall. Lumberjacks rely almost wholly on saw cuts and wedges to influence a tree's fall. But in residential felling, rope can add a margin of safety.

Rope. There are many sad stories of homeowners who relied too heavily on rope. In one case, two fellows notched a large birch to fall away from its natural lean, which was toward a house. They then attempted to pull the tree over with a rope attached to a car. The car pulled, the rope broke, and the tree crushed the house.

This tree should have been limbed first and then lifted toward vertical by means of wedges in the back cut. Several strong ropes should have been used during the pull—one rope for the pull itself and the two others to prevent the tree's falling backward. Car pulls are risky because they don't allow you enough feel for rope stress. Under ideal conditions, two men pulling tug-o'-war fashion will be enough. Otherwise, a lone man can operate a block and tackle or else a ratcheted come-along. I use a come-along with a 1000-pound recommended maximum limit. Then I never lever it over what I estimate to be a few hundred pounds. This way I ensure that the come-along won't overstress the pull rope.

Because good rope is expensive, many people improvise by splicing odd lengths. This is often a mistake. Especially suspect ropes are those that have mildewed in cellars and those that show abrasion. Knots that join ropes of different diameters tend to slip. Here the sheet bend probably serves best. Also, beware of

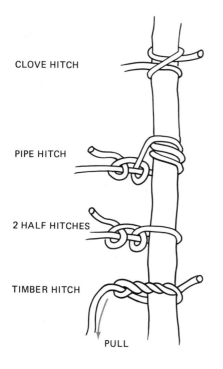

CLOVE HITCH

PIPE HITCH

2 HALF HITCHES

TIMBER HITCH

PULL

These knots are often used interchangeably for similar purposes. But each has its advantages and disadvantages. The timber hitch is used for dragging logs; it absolutely won't slip, but it can't be tied while there's strain on the rope. Two half hitches can be tied while there's strain on the rope, but they can be difficult to untie if they've carried heavy loads. The pipe hitch is just two half hitches with a few preliminary wraps; it's used to pull pipe from the earth and it works well for lowering slippery-barked limbs from tree tops. The clove hitch is quick to tie and untie, and it's reliable under steady tension.

BOWLINE

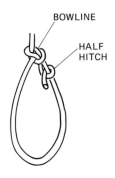

BOWLINE

HALF HITCH

The bowline (pronounced *bo-lin*) is the basic knot of all professionals who must trust a rope with their life. It won't slip or jam, and it's easy to untie. The first drawing shows a basic bowline, and the second shows an extra hitch that shortens the end and adds security.

BOWLINE

BOWLINE WITH EXTRA HITCH

The sheet bend is used to join ropes and is especially reliable for ropes of different thicknesses. For heavy loads, a double sheet bend is advised. Its free ends can then make a few half hitches for added security.

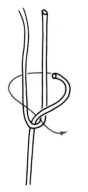

SHEET BEND   DOUBLE SHEET BEND

BACKWALL HITCH

The backwall hitch stresses the rope less than a simple loop would, and it won't jump off the hook during a sudden jerk.

WITH 2 HALF HITCHES

Left, a ½-inch manila loop is joined to ⅜-inch nylon by means of a simple sheet bend and two half hitches. Right, a ⅜-inch kermantle mountain-climbing rope with bowline suspends a D-shaped carabiner.

an unwanted overhand knot that has somehow worked its way into the rope. Under stress, it will bind upon itself, shearing its own fibers. Such a knot can reduce a rope's strength by half.

Manila has a hard lay, meaning it is stiff enough to resist much kinking on its own. It resists stretching. It's the favored rope for tree removal. Top-grade, ½-inch manila costs over 25¢ a foot. This rope is abrasion-resistant and treated to prevent rot. It is normally advertised to offer a breaking strength of over 2000 pounds. But you're smart not to stress a new rope beyond about one-seventh of its advertised breaking strength. Older ropes deserve even gentler consideration.

Polypropylene ropes are advertised as 50 percent stronger than manila of the same thickness yet cost a little less. Nylon is generally advertised to possess about double the breaking strength of same-thickness manila, but it often costs half again as much.

All synthetic ropes can be damaged by heat or petroleum products. And like manila, all can be abraded dangerously by grit and stones that become embedded in the strands. All synthetic ropes tend to stretch more than manila. This makes them undesirable for pulling—and even dangerous. About the only advantage of synthetics in tree work is their stretch when used to drop and lower light limbs. The stretch affords some shock absorption. But if synthetics are used on heavy limbs that drop some distance, the resiliency of the rope causes the limb to bounce in the air, offering added challenges for the ground man and making the work more precarious for anyone in the tree.

Rope caution. When rope is relied on to overpower a stubborn tree, the tree may wind up the winner. In felling, rope should be used to prevent a tree's "sitting back" on the back cut and falling in the wrong direction. Heavy rope or chain should be used to anchor the bottom of the falling trunk enough to prevent its damaging property, should it jump off the stump. Rope can also be used to exert a few hundred pounds of pull to start the tree falling in the right direction. From there, gravity and the fulcrum provided by the hinge wood should control the fall. If you use tremendously heavy rope and heavy-duty

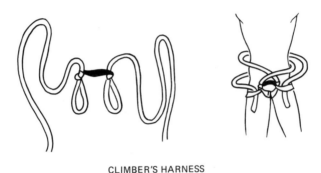

CLIMBER'S HARNESS

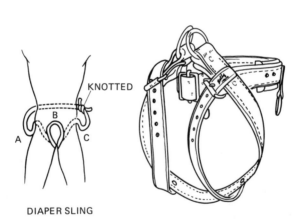

DIAPER SLING

TREE-CLIMBING
HARNESS

Safety belts should give support around the thighs as well as around the waist. The diaper sling can be rigged from 8–10 feet of mountain-climbing rope or webbing; double carabiners are used to hold A, B, and C together. The climber's harness requires 20–30 feet of webbing; its free ends are wrapped around the waist and then tied off. A commercial tree-climbing harness costs only a little more than quality webbing.

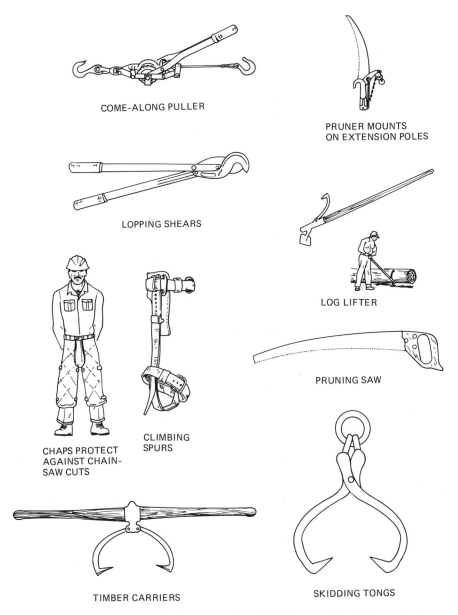

COME-ALONG PULLER

PRUNER MOUNTS
ON EXTENSION POLES

LOPPING SHEARS

LOG LIFTER

CHAPS PROTECT
AGAINST CHAIN-
SAW CUTS

CLIMBING
SPURS

PRUNING SAW

TIMBER CARRIERS

SKIDDING TONGS

Here are a few of the items for tree work available from mail-order houses.

pullers, you may wind up forcing a tall split up the trunk just as the tree begins to fall. This usually results in a stump with high spires of wood fiber jutting up, which loggers call a barber's chair. That's only embarrassing, unless the splitting lets the trunk twist or otherwise jump the stump. A large leaning tree can jump backward 20 feet and cause a lot of grief.

Other gear. Forestry-supply catalogs listed in the Appendix offer wide assortments of gear for tree work. For climbing there are safety belts of many styles, leg irons, and ladders strange and conventional. There is hardware for working felled logs. There is protective gear that will cover every part of you—from head to toe, as well as inside your ears.

The gear you'll need depends on the amount of tree work you do and the problems you want to tackle. It's safest never to attempt a job you lack the the equipment for. There isn't space here to discuss all the possibilities. But I feel compelled to talk about two items: safey belts and climbing spurs.

Safety belts are essential if you do any climbing, even if you never leave the top rungs of a ladder. Lashing a strong rope around your waist and tying a loose end to a limb is better than nothing, but a short fall with such a rope belt can snap your back, break your ribs, or scorch your armpits. Strong but simple belts like those used by linemen and window washers are an improvement on a simple rope, but they can hurt you much the way the rope belt can. The best tree belts have loops for each thigh that can be cinched tight through the crotch and around your rump. These belts are designed to catch you, in a fall, in a sitting position. They also allow you to lean out from limbs more safely and gain better leverage on the work. Good belts for tree work cost over $40. If you plan to do a lot of climbing, they are worth the money.

As alternatives though, a mountain climbing diaper sling or a harness works well. In a short fall, a diaper sling won't be as comfortable as a harness, but it will arrest your fall with far less chance of causing serious injury than would a rope merely tied around your waist.

Climbing spurs are similar to the leg irons used by linemen on telephone poles, except that the spur (gaff) is 3½ inches long. Linemen's spurs are only 2 inches long. The longer spurs allow better penetration through bark and into sapwood. But spurs are dangerous! If a spur breaks loose from the tree unexpectedly, your safety belt will slam your crotch into the tree and then maybe drop you in short jerky increments until either you or the belt hangs up on something. Then too, a spur can pierce the inside of an opposite foot or ankle badly enough to cripple you for life, not to mention the blood it can let.

If you eventually do a lot of tree work, you'll probably have need of climbing spurs, and if you have to limb very large trees, spurs may offer your only means of reaching the necessary heights. Myself, I plan to decline all job opportunities that would make me strap on climbing spurs. If I can't handle a job with an extension ladder and some fairly easy free climbing, I won't bid.

BUCKING FELLED TREES. Many weekend loggers are so relieved once their trees are felled that they plunge into bucking with light-hearted recklessness. It's true that a felled tree can no longer fall on bystanders or a house—and that no one can fall out of the tree now. But bucking can be very dangerous work.

The chief hazards from the felled tree itself involve wood fibers under stress

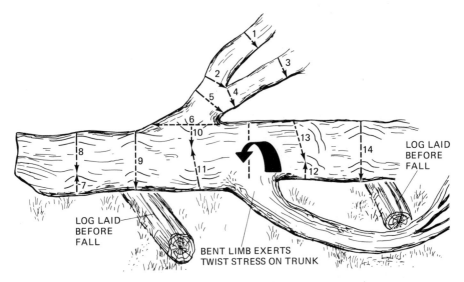

This diagram shows the best way to buck and limb a felled tree. Generally it's best to buck a tree from bottom to top.

Cuts 1–5: Simple overbucks to stove length.

Cuts 5, 6: Crotch sections cut shorter than stove length to make splitting easier.

Cuts 7, 8: Compressed wood underbucked, tensioned wood overbucked.

Cut 9: Tensioned wood overbucked.

Cuts 10, 11: Compressed wood overbucked. Tensioned wood underbucked at an angle to allow short end section to fall without pinching the saw bar.

Cuts 12, 13: Compressed wood underbucked. Angled overbuck lets twisting stress from bent limb force trunk down.

Cut 14: Tensioned wood overbucked to provide working room near bent limb.

and rolling trunks. A trunk section suspended between two supporting limbs will be under a sagging sort of stress, with the bottom side pulled tense and the top side compressed. A trunk section over a support will evidence an upward arcing stress, with the bottom fibers compressed and the top fibers pulled tense. Limbs supporting the trunk want to break in the direction they arc and may evidence twisting stresses in addition to simple arcing.

Stresses can make sawing difficult, because fibers under compression will tend to pinch a saw blade. Those under tension will sever easily—too easily! In most cases, highly tensed fibers will result in severed ends that lurch free or even whip free. A whipping limb can tear your head off or otherwise inflict bruises, and it can whip a chain-saw bar as viciously as if the saw had kicked back. Then again, if you cut a supporting limb on the downhill side of a trunk, the trunk can roll on you.

Besides the possibility of a chain saw's being whipped at you by a limb, there's increased hazard of kickback during bucking. This results from accidental touching of the nose of the bar against limbs or the ground. Also, when most people buck a felled tree, they wind up walking and vaulting with a running chain saw. If you stumble onto the teeth of a motionless chain, you can receive nasty gashes. If the chain is moving—oh brother! Trench-like cuts ½ inch wide through flesh and bone heal very slowly, even if you get prompt medical attention.

Residential trees often have served as supports for tree houses, bird houses, clotheslines, and plaques. This means nails and large spikes that may be visible outside the bark or that may be buried under layers of tree rings. Surface nails should normally be removed before you fell the tree. But buried nails will remain hidden. Remember that nails, like branch stubs, do not rise up the tree as it grows; they remain at their original elevations. Sometimes owners will be able to advise you of approximate locations of old buried nails—often not. A nail can ruin the teeth on a saw chain. It can break a chain too, increasing the hazard of your being cut by flying chain ends. Nails and chain saws are bad business.

Hand tools for bucking. Because of chain-saw hazards associated with felled trees, I recommend using hand tools as long as they are adequate for the tasks. Heavy-duty lopping shears can make quick, safe work of limbs up to 1½ inches in diameter.

A sharp ax and a bow saw or a pruning saw can limb most conifers with remarkable speed. Here the comparatively small limbs and normal symmetry of conifers simplifies the tasks. On conifers, though, keep the trunk between you and the ax if possible, and work from the trunk toward the top. Make sure the ax has enough clearance as you swing it. If clearance is limited or if there's danger from ricochet, switch to a hand saw.

The branching patterns of most hardwoods usually make ax work difficult. Here a bow saw can, and should, handle all limbs under 3 inches in diameter, except for those taken with the lopping shears. Depending on the tree's lie, there may be advantages in bucking large trunk sections free to lighten the load on supporting limbs.

APPENDIX: Supply Sources

Appendix 1: General Mail Order Sources

The Ben Meadows Co.
3589 Broad St.
Atlanta (Chamblee), GA
30366

A wide range of tools and protective gear for logging and tree work. Ask for info on specific types of items.

Garden Way Marketplaces
509 Westport Ave.
Norwalk, CT 06851

Basic back-to-the-earth equipment ranging from seeds and dehydrators to stoves and wheel-barrow carts.

TSI Company
PO Box 151
25 Ironia Rd.
Flanders, NJ 07836

A wide range of tools and protective gear for logging and tree work.

Appendix 2: Sources for Power Splitters

COMPANY	SCREW-TYPE	WEDGE-TYPE	COMMENT
Albright Welding Corp. Jeffersonville, VT 05464		self-powered and tractor-PTO-powered models	called L.O. Balls Woodsplitters; models taking logs up to 4 ft. long
Amerind-MacKissic, Inc. PO Box 111 Parker Ford, PA 19457		self-powered	called Mighty Macs; two models taking logs to 19 and 26 inches; also leaf and twig shredder/chippers

Sources for Power Splitters (*continued*)

COMPANY	SCREW-TYPE	WEDGE-TYPE	COMMENT
Didier Manufacturing Co. 8630 Industrial Dr. Franksville, WI 53126		self-powered and tractor-PTO-powered models	six models, the largest taking logs 4 ft. long; also other farm and garden machinery
Household Wood-Splitters PO Box 143 Jeffersonville, VT 05464		self-powered and tractor-PTO-powered models	five models, the largest taking logs to 26 inches
Nortech Corp. 300 Greenwood Ave. Midland Park, NJ 07432	self-powered and tractor-PTO-powered models		called the Screw-Wedge
Piqua Engineering, Inc. PO Box 605 Piqua, OH 45356		self-powered	about 20 models taking logs ranging from 26 inches to 9½ feet; also brush chippers
Richard Pokrandt Mfg. RD 3, Box 182 Tamaqua, PA 18252		self-powered and tractor-PTO-powered models	also PTO cordwood saw and saw-splitter combination
Taos Equipment Manufacturers Box 1565 Taos, NM 87571	car-wheel hub and tractor-PTO powered models		called The Stickler
Thackery Co. 1879 Frebis Ave. Columbus, OH 43206	car, truck, or garden tractor wheel-hub powered		called the Unicorn
Thrust Mfg., Inc. 6901 S. Yosemite St. Englewood, CO 80110		self-powered	

Appendix 3: Sources for Wood-burning Equipment

The following sources were alive and well as this book went to press. All indicated a willingness to send sales literature in response to written requests. Those companies requiring fees to cover postage and printing are noted in the "Comment" column.

COMPANY	STOVES	FIREPLACES	FURNACES/BOILERS	ACCESSORIES	COMMENT
Abundant Life Farm, Inc. PO Box 63 Lochmere, NH 03252	radiating				
Aglow Heat-X-Changer PO Box 10427 Eugene, OR 97401				fireplace heat extractor	glass doors with grate and blower
American Stovalator Route 7 Arlington, VT 05250	stove-like box insert for fireplace				glass doors give fireplace look
Aquappliances, Inc. 135 Sunshine Lane San Marcos, CA 92069				fireplace heat extractor	glass doors with grate and blower
Atlanta Stove Works PO Box 5254 Atlanta, GA 30307	radiating, circulating, cooking	freestanding			
Bellway Manufacturing Grafton, VT 05146	radiating		furnaces and boilers	screw-type wood splitter	Splitter mounts on car wheel
Blazing Showers PO Box 327 Point Arena, CA 95468				hot-water heat extractors for stovepipe and firebox	plus book on plumbing connections
Bow & Arrow Imports 14 Arrow St. Cambridge, MA 12138	radiating, circulating, cooking				importer of French coal/wood stoves
Carlson Mechanical Contractors, Inc. PO Box 242 Prentice, WI 54556			boiler		
Cawley LeMay Stove Co. Box 431, RD 1 Barto, PA 19504	radiating cooking stoves				
Charmaster Products 2307 Highway 2 W. Grand Rapids, MN 55744		furnace/fireplace combo	wood and multi-fuel furnaces		

Sources for Woodburning Equipment (*continued*)

COMPANY	STOVES	FIREPLACES	FURNACES/ BOILERS	ACCESSORIES	COMMENT
Chimney Heat Reclaimer Corp. 53 Railroad Ave. Southington, CT 06489	circulating			stovepipe heat extractor	
Columbia Automatic Heater Co. PO Box 6472 (Suite 513) Lakeshore Complex Columbia, SC 29260	radiating, circulating				
Combo Furnace Co. 1707 W. 4th St. Grand Rapids, MN 55744			wood and multi-fuel furnaces and boilers		
C & D Distributors, Inc. PO Box 766 Old Saybrook, CT 06475	radiating			broiler, poker, chimney sweeper	stove installs in old fireplace
Damsite Dynamite Stove Co. RD 3 Montpelier, VT 05602	radiating, cooking		furnace		optional water tanks for hookup to home hot water
Dual Fuel Products, Inc. Box 514 Simcoe, Ontario	radiating		multi-fuel furnace		
Duo-matic Division Manoir International 2413 Bond St. Park Forest South IL 60466			wood and multi-fuel furnaces		
Edmund Scientific Co. 7782 Edscorp Bldg. Barrington, NJ 08007				stovepipe heat extractors	
Energy Associates PO Box 524, Dept. 2 Old Saybrook, CT 06475				metal sawbuck, log rack, splitting maul	Include 25¢ postage
Fisher Stoves, Inc. Rt. 3, River Rd. Bow, NH 03301	radiating	freestanding			

Company	Stove type	Configuration	Furnace/boiler	Accessories	Notes
Greenbriar Products, Inc. Box 473 Spring Green, WI 53588	fireplace stove				
HDI Importers Schoolhouse Farm Etna, NH 03750	radiating, ceramic tile and tile-lined circulating				importer of custom-built tile stoves from Austria; costly
Heatilator Fireplace 1922 W. Saunders St. Mt. Pleasant, IA 52641		circulating, free-standing			
Hinckley Foundry & Marine 13 Water St. Newmarket, NH 03857	radiating			ash apron for Jøtul stoves	
Hunter Enterprises PO Box 400 Orillia, Ontario CANADA L3V 6K1	radiating, circulating		wood and multi-fuel furnaces		
Hydroheat Div. Ridgway Steel Fabricators, Inc. PO Box 382 Ridgway, PA 15853	water-circulating	water-circulating grates			
Integrated Thermal Systems 379 State St. Portsmouth, NH 03801			wood and multi-fuel boilers		distributes Tasso (wood) and HS Tarm (multi-fuel)
Kickapoo Stove Works, Ltd. Box 127-58 LaFarge, WI 54639	radiating	freestanding	furnace	chimney brushes, stovepipe ovens, cookware	
Kristia Associates PO Box 1118 Portland, ME 04104	radiating	freestanding			importer of Jøtul stoves
Locke Stove Co. 114 W. 11th St. Kansas City, MO 64105	radiating, circulating, and barrel-stove kits				
Longwood Furnace Corp. Rt. 1, Box 223 Gallatin, MO 64640			multi-fuel furnace		

Sources for Woodburning Equipment (*continued*)

COMPANY	STOVES	FIREPLACES	FURNACES/BOILERS	ACCESSORIES	COMMENT
Louisville Tin & Stove Co. PO Box 1079 Louisville, KY 40201	radiating drum-style			stovepipe, connections, stoveboards, ash scuttle and shovel	
Lyndale Manufacturing Co., Inc. 1309 North Hills Blvd. (Suite 207) North Little Rock, AR 72116			multi-fuel furnaces		
Majestic Co. Huntington, IN 46750		freestanding, prefab			
Malleable Iron Range Co. 715 N. Spring St. (Dept. PSBC78) Beaver Dam, WI 53916	radiating, wood and multi-fuel kitchen ranges	freestanding	furnace	cast iron grates and kettles	
Malm Fireplaces, Inc. 368 Yolanda Ave. Santa Rosa, CA 95404	radiating	freestanding, prefab circulating			
Marathon Heater Co., Inc. Box 265, RD 2 Marathon, NY 13803			multi-fuel boiler		
Metal Concepts, Inc. 15800 9th NE Seattle, WA 98155				fireplace heat extractor	
Mohawk Industries, Inc. 173 Howland Ave. Adams, MA 01220	radiating, cooking				
Old Country Appliances PO Box 330 Vacaville, CA 95688	cooking, radiating, circulating				importer of Tirolia wood/coal ranges; distributor of Reginald kit stove
Pioneer Lamps & Stoves 71 Yesler Way Seattle, WA 98104	old-time wood range				
Preway Wisconsin Rapids WI 54404		freestanding, prefab		grates, blowers, prefab chimneys	

192

Company			furnaces, boilers		
Ram Forge Co. Brooks, ME 04921	radiating				
Riteway PO Box 6 Harrisonburg, VA 22801	radiating, circulating		wood and multi-fuel furnaces and boilers	water heaters and jackets for stoves	
Self Sufficiency Products One Appletree Square Minneapolis, MN 55420	radiating		furnaces		
SEVCA Stove Works, Inc. Box 396 Bellows Falls, VT 05101	radiating				Include 25¢ postage
Shenandoah Mfg. Co. Inc. PO Box 839 Harrisonburg, VA 22801	radiating, circulating			fire-grate heat extractor	
John P. Smith Co. 174 Cedar St. Branford, CT 06405				fireplace screens, log rack	
Solar Sauna Box 466 Hollis, NH 03049	sauna stoves			sauna water tank	
Solar Wood Energy Corp. E. Lebanon, ME 04027			furnace with domestic hot-water coils		
Southport Stoves, Inc. 248 Tolland St. E. Hartford, CT 06108	radiating	fireplace stoves			importer of Morso stoves
The StoveWorks Box 172 Marlboro, VT 05344	radiating, circulating		furnace		Include 50¢ postage
Sturges Heat Recovery PO Box 397 Stone Ridge, NY 12484		prefab		heat extractor units for stoves and fireplaces	
Superior Fireplace Co. 4325 Artesia Ave. Fullerton, CA 92633		prefab			
Torrid Mfg. Co., Inc. 1248 Poplar Place S. Seattle, WA 98144	radiating	freestanding, prefab		stovepipe oven, heat extractor	

also sells plans for integrated greenhouse, solarium, sauna

Sources for Woodburning Equipment (*continued*)

COMPANY	STOVES	FIREPLACES	FURNACES/BOILERS	ACCESSORIES	COMMENT
United States Stove Co. PO Box 151 South Pittsburg, TN 37380	radiating, circulating, cooking			fire screen, wood rack, bean pot and hook	
Vermont Castings, Inc. Box 126, Prince St. Randolph, VT 05060	radiating				Include $1 postage
Vermont Iron Stove Works at the Bobbin Mill Box 252B Warren, VT 05674	radiating				250-lb. barrel-shaped extravaganza; "The Elm"
Vermont Woodstove Co. PO Box 1016 Bennington, VT 05201	DownDrafter with blowers			water coils, manifold for ductwork, stack thermometer	
Washington Stove Works PO Box 687 Everett, WA 98206	radiating, cooking	freestanding, prefab		barrel-stove kit	
WESCO Products Bellfast, ME 04915	radiating, circulating		furnaces, boilers		
Whittier Steel & Mfg. 10725 S. Painter Ave. Santa Fe Springs, CA 90670	radiating	prefab		insulated fluepipe, spark-arrester cap	
Wilson Industries, Inc. 2296 Wycliff St. St. Paul, MN 55114			wood and multi-fuel furnaces		
Woodburning Specialties PO Box 5 North Marshfield, MA 02059			wood and multi-fuel furnaces		importer of Hunter (Canada)
Yankee Woodstoves Box 7, Bible Hill Rd. Bennington, NH 03442	radiating barrel-stove kits				

Index